Pharmacology – Research, Safety Testing and Regulation

Human Serum Albumin

Structure, Binding and Activity

PHARMACOLOGY – RESEARCH, SAFETY TESTING AND REGULATION

Additional books and e-books in this series can be found on Nova's website under the Series tab.

PHARMACOLOGY – RESEARCH, SAFETY TESTING AND REGULATION

HUMAN SERUM ALBUMIN

STRUCTURE, BINDING AND ACTIVITY

DIANNE COHEN
EDITOR

Copyright © 2019 by Nova Science Publishers, Inc.

All rights reserved. No part of this book may be reproduced, stored in a retrieval system or transmitted in any form or by any means: electronic, electrostatic, magnetic, tape, mechanical photocopying, recording or otherwise without the written permission of the Publisher.

We have partnered with Copyright Clearance Center to make it easy for you to obtain permissions to reuse content from this publication. Simply navigate to this publication's page on Nova's website and locate the "Get Permission" button below the title description. This button is linked directly to the title's permission page on copyright.com. Alternatively, you can visit copyright.com and search by title, ISBN, or ISSN.

For further questions about using the service on copyright.com, please contact:
Copyright Clearance Center
Phone: +1-(978) 750-8400   Fax: +1-(978) 750-4470   E-mail: info@copyright.com.

NOTICE TO THE READER

The Publisher has taken reasonable care in the preparation of this book, but makes no expressed or implied warranty of any kind and assumes no responsibility for any errors or omissions. No liability is assumed for incidental or consequential damages in connection with or arising out of information contained in this book. The Publisher shall not be liable for any special, consequential, or exemplary damages resulting, in whole or in part, from the readers' use of, or reliance upon, this material. Any parts of this book based on government reports are so indicated and copyright is claimed for those parts to the extent applicable to compilations of such works.

Independent verification should be sought for any data, advice or recommendations contained in this book. In addition, no responsibility is assumed by the publisher for any injury and/or damage to persons or property arising from any methods, products, instructions, ideas or otherwise contained in this publication.

This publication is designed to provide accurate and authoritative information with regard to the subject matter covered herein. It is sold with the clear understanding that the Publisher is not engaged in rendering legal or any other professional services. If legal or any other expert assistance is required, the services of a competent person should be sought. FROM A DECLARATION OF PARTICIPANTS JOINTLY ADOPTED BY A COMMITTEE OF THE AMERICAN BAR ASSOCIATION AND A COMMITTEE OF PUBLISHERS.

Additional color graphics may be available in the e-book version of this book.

Library of Congress Cataloging-in-Publication Data

ISBN: 978-1-53614-787-2

Published by Nova Science Publishers, Inc. † New York

Contents

Preface		**vii**
Chapter 1	Human Serum Albumin Structure and Intracellular Content *Sabina Galiniak, David Aebisher and Dorota Bartusik-Aebisher*	**1**
Chapter 2	Human Serum Albumin and Antioxidant Properties *David Aebisher, Dorota Bartusik-Aebisher and Łukasz Ożóg*	**19**
Chapter 3	Evaluation of Photodynamic Agent Activity Using Human Serum Albumin *Kazutaka Hirakawa*	**37**
Chapter 4	Magnetic Resonance Imaging and Spectroscopy for the Determination of Human Serum Albumin *Dorota Bartusik-Aebisher, David Aebisher and Adrian Truszkiewicz*	**69**

Chapter 5	The Influence of Fluorine-19 Drug Labeling on Binding to Human Serum Albumin *Dorota Bartusik-Aebisher, David Aebisher and Zuzanna Bober*	**79**
Chapter 6	Tyrosine Detection in Microalbuminuria *Dorota Bartusik-Aebisher, David Aebisher, Sabina Galiniak, Łukasz Ożóg and Małgorzata Marć*	**91**
Index		**101**

PREFACE

Human serum albumin is found in the intravascular and extracellular space and is the main protein of human blood plasma which binds water, cations (such as Ca2+, Na+, K+), fatty acids, hormones, bilirubin, thyroxin (T4) and pharmaceuticals. Structurally, the serum albumins are similar, each domain containing five or six internal disulfide bonds. In the opening chapter of Human Serum Albumin: Structure, Binding and Activity the authors review the structure, content and binding of Human Serum Albumin.

Then, the role of albumin in free radical trapping activities and as an oxyradical scavenger is described. A discussion of recent advances in the use of the antioxidant properties of human serum albumin to make drugs detectable in vivo is also presented.

Human serum albumin has one tryptophan residue and shows a characteristic fluorescence of around 350 nm under ultraviolet irradiation. Because tryptophan is easily oxidized by reactive oxygen species and/or photoexcited molecules through electron transfer (leading to fluorescence diminishment) a fluorometry of this tryptophan residue is a useful tool to evaluate oxidation. In light of these characteristics, the authors examine the photosensitizing activity of organic photosensitizers, including porphyrins and phenothiazine dyes.

The use of magnetic resonance imaging and spectroscopy for the determination of human serum albumin structure, drug binding and in vivo activity is explored, in addition to drug modifications using human serum albumin.

Following this, this compilation studies the major approaches for the characterization of human serum albumin as a fluorinated drug delivery agent and fluorinated albumin influence on drug binding. Synthesis and characterization of fluorinated conjugates of albumin and adsorbed human serum albumin on surfaces containing CF3 are also discussed.

The concluding study investigates possible similarities and differences in albumin concentration and the presence of tyrosine in urine from a population of healthy and microalbuminuria dependent women. The assessment of subtle changes in albumin concentration, the primary macromolecular component of urine, is critical for the diagnosis of early stage albuminuria, one of the major complications in nephropathy.

Chapter 1 - Human serum albumin (HSA) is found in the intravascular and extracellular space and is the main protein of human blood plasma. Human serum albumin binds water, cations (such as Ca^{2+}, Na^+, K^+), fatty acids, hormones, bilirubin, thyroxin (T4) and pharmaceuticals. Structurally, the serum albumins are similar, each domain containing five or six internal disulfide bonds. In this chapter the authors review the structure, content and binding of HSA.

Chapter 2 - Human serum albumin represents an important circulating antioxidant. This review brings together major recent insights on human serum albumin antioxidant properties. Here the authors describe the role of albumin in free radical trapping activities and as an oxyradical scavenger. A discussion of recent advances of the use of human serum albumin antioxidant properties to make drugs detectable in vivo is also presented.

Chapter 3 - Human serum albumin (HSA) is a water-soluble protein that can hydrophobically interact with organic molecules. HSA has one tryptophan residue and shows a characteristic fluorescence at around 350 nm under ultraviolet irradiation. Because tryptophan is easily oxidized by reactive oxygen species and/or photoexcited molecules through electron transfer, leading to fluorescence diminishment, a fluorometry of the HSA tryptophan residue is a useful tool to evaluate HSA oxidation. In light of these HSA characteristics, the authors examined the photosensitizing activity of organic photosensitizers, including porphyrins and phenothiazine dyes. The purpose of this study using HSA is the development of agents (photosensitizers) for photodynamic therapy (PDT), which is a less invasive treatment of cancer and infections. Administered photosensitizers undergo visible light irradiation, resulting in the photochemical oxidation of targeting biomacromolecules to kill cancer cells and pathogenic bacteria. Therefore, photosensitized protein-damaging activity is an important property of photodynamic agents. Absorption spectral change can be used to examine the interaction between HSA and photosensitizers. Analysis of the relationship between the absorbance of photosensitizers in a visible light region and the concentration of HSA gives us the association constant and binding ratio. Fluorescence lifetime measurement of the tryptophan residue of HSA is a simple method to estimate the binding distance with photosensitizing molecules. After visible light irradiation of the sample solution consisting of HSA and photosensitizers, damaged contents of HSA can be estimated by measuring the intrinsic fluorescence intensity of the tryptophan residue. Using this simple system, the activity of photosensitizers and the damaging mechanisms can be elucidated. In this chapter, recent examples of the applications of HSA to study PDT photosensitizers are introduced.

Chapter 4 - The aim of this chapter is to review the use of Magnetic Resonance Imaging and Spectroscopy for the determination of human serum albumin structure, drug binding and in vivo activity. PubMed, Embase, the Cochrane Library, Elsevier, Wiley, and Ovid were searched for randomized controlled trials and prospective studies. Also, the authors discuss studies which describe drug modifications using Human Serum Albumin.

Chapter 5 - The authors review the major approaches for the characterization of human serum albumin as a fluorinated drug delivery agent and fluorinated albumin influence on drug binding. Synthesis and characterization of fluorinated conjugates of albumin, adsorbed human serum albumin on surfaces containing CF_3 are also discussed. The databases such as PubMed, ScienceDirect and Springer were utilized to search the literature for relevant articles.

Chapter 6 - The assessment of subtle changes in albumin concentration, the primary macromolecular component of urine, is critical for the diagnosis of early stage albuminuria, one of the major complications in nephropathy. The aim of this study was to investigate possible similarities or differences in albumin concentration and the presence of tyrosine in urine from a population of healthy and microalbuminuria dependent women.

In: Human Serum Albumin
Editor: Dianne Cohen

ISBN: 978-1-53614-787-2
© 2019 Nova Science Publishers, Inc.

Chapter 1

HUMAN SERUM ALBUMIN STRUCTURE AND INTRACELLULAR CONTENT

Sabina Galiniak, David Aebisher and Dorota Bartusik-Aebisher[*]

Faculty of Medicine, University of Rzeszów, Rzeszów, Poland

ABSTRACT

Human serum albumin (HSA) is found in the intravascular and *extracellular* space and is the main protein of human blood plasma. *Human serum albumin* binds water, cations (such as Ca^{2+}, Na^+, K^+), fatty acids, hormones, bilirubin, thyroxin (T4) and pharmaceuticals. Structurally, the serum albumins are similar, each domain containing five or six internal disulfide bonds. In this chapter we review, the structure, content and binding of HSA.

Keywords: human serum albumin, hyperalbuminemia, hypoalbuminemia

---

[*] Corresponding Author Email: dbartusik-aebisher@ur.edu.pl.

Human serum albumin (HSA) is the most abundant protein of blood plasma and constitutes about 60% of the total blood protein and 3% of total body protein. It has very important physiological and biochemical functions. Normal concentrations of HSA in blood serum is about 35-50 g/L (Choi et al. 2004). HSA values below the reference ranges (hypoalbuminemia <35 g/L or significant hypoalbuminemia <25g/L) are associated with several diseases including cirrhosis, malnutrition, nephrotic syndrome and sepsis (Gatta et al. 2012), whereas hyperalbuminemia (above 50g/L) is associated with high-protein diets and acute dehydration (Fagan & Oexmann 1987). Moreover, low albumin levels in hospitalized patients are correlated to increased short- and long-term mortality (Akirov et al. 2017), while high HSA level is an indicator of a beneficial response to immunotherapy in autoimmune encephalitis (Jang et al. 2018). Data has shown that hipo- and hyperalbuminemia is related to, respectively, the decrease and increase in serum anion gap which is useful in the interpretation of acid-base disorders (Feldman et al. 2005). It is believed that human serum albumin is a reliable marker of many disease states, including neoplastic diseases, rheumatoid arthritis, ischemia, and post-menopausal obesity. Administration of albumin is used as a clinical treatment of various diseases such as hypovolemia, shock, burns, surgical blood loss, trauma, hemorrhage, cardiopulmonary bypass, acute respiratory distress syndrome, haemodialysis, acute liver failure or chronic liver disease (Fanali et al. 2012).

In view of the fact that HSA consists of large number of acidic and basic amino acid residues, HSA is a highly soluble protein that also occurs in secreted and excreted liquids from the human body including urine, milk, sweat, tears and saliva. The content of HSA in sweat, saliva and tears is about 0.1-0.5 g/L. Lower amounts of HSA are found in cerebrospinal, synovial and lung bronchoalveolar lining fluids (Peters 1996).

Albumin is a multi-functional protein with antioxidant, immune-modulatory, and detoxification functions. It is responsible for 80% of the plasma oncotic pressure, due to high concentration in plasma and its negative charge. HSA is also involved in binding and transporting various molecules in circulation. Likewise, the protein is involved in stabilisation of endothelial cells and the maintenance of capillary permeability (Caraceni et al. 2013).

The first reports on human serum albumin research appeared in the 19th century. The three-dimensional structure of human serum albumin was revealed in 1989 at a low resolution of 6.0 Å (Carter et al. 1989), and at a high resolution of 2.5 Å by Sugoi and co-workers in 1999 (Sugoi et al. 1999). The ellipsoid shape of the albumin molecule is close to heart shape, with approximate dimensions of 7.5 × 6.5 × 4.0 nm (Erickson 2009). HSA is composed of a single polypeptide chain with 585 amino acids with molecular mass of 66.5 kDa. Complete sequences of HSA were discovered and published in 1975 by Melon et al. (Meloun et al. 1975). The main amino acid of albumin is glutamic acid, which accounts for 15%, equivalent to 82 residues. In the protein structure, other abundant amino acids residues include alanine, leucine and lysine; HSA has a low content of methionine, glycine and isoleucine. HSA contains only a single tryptophan residue at position 214 (Peters 1996).

The gene for HSA is located in chromosome 4 at position q13.3 and comprises 16,961 nucleotides split into 15 exons by sequences which are symmetrically placed within the three domains of albumin (Hawkins and Dugaiczyk 1982; Minghetti et al. 1986).

Albumin is a molecule mostly composed of an α-helix with three domains named Domain I, II and III which are structurally similar. Precisely, the HSA secondary structure consists of about 67% α-helix and β-sheet elements (Peters 1996). Studies conducted by Bramanti and Benedetti has shown that raising the temperature to 70° and 90° causes a decrease in the α-helix content to 34% and 2%, respectively (Bramanti and Benedetti 1996). Value of pH also has impact on the structure of HSA. The protein occurs in an extended conformation under very acidic conditions (lower than 2.7), while at pH above 8.0, HSA structure changes to a basic form with reduction of α-helix content (Fanali et al. 2012).

Domain I consist of 195 amino acid residues, while domain II and III comprise 188 and 202, respectively. Each of domain has three loops named sub-domains – two long labelled A and one shorter labelled B. Sub-domain A consist of four α-helices which are surrounded by two short α-helices while sub-domain B is a complex of four α-helices. Both sub-domains are connected by extended loop and a pseudo-binary axis that gives HSA its heart-shape (Tarhoni et al. 2008). Despite similar domains, they have different properties i.e., the ability to bind ligands and other functionalities. HSA has two main specific binding sites named Sudlow's site I found in subdomain IIA and Sudlow's site II located in subdomain IIIA. Several additional bindings' sites have also been described (Fasano et al. 2005; Wang et al. 2015).

The structure of albumin is stabilized by 17 disulphide bridges formed between cysteine residues which allow significant alternation of shape and size of HSA in response to various factors. In the protein tertiary structure, certain areas overlap as a result of the helical continuation. Albumin has a single thiol at cysteine-34, which constitutes about 80% of the total thiols in plasma and is a plasma scavenger of oxygen and nitrogen reactive species. Cysteine-34 is located in subdomain IA, hence its importance in binding ligands affinity.

Albumin is synthesized by hepatocytes with an efficiency of ~13.9 g of HSA per day. It is known that newly synthesized albumin, as preproalbumin, is transported from the rough endoplasmic reticulum to the smooth-surfaced endoplasmic reticulum where modifications occur. From the endoplasmic reticulum lumen, proalbumin without N-terminal peptide is transported to the trans-Golgi network. From the Golgi apparatus, the albumin molecule is sub fractionated together with very low-density lipoproteins into secretory vesicles and transported into the blood stream (Gudehithlu et al. 2004).

Hormones, such as insulin or cortisol, stimulate synthesis of HSA in hepatocytes, while proinflammatory substances are inhibitors of the synthesis process (Caraceni et al. 2013).

In serum, albumin has a half-life of approximately 19-20 days under normal conditions (Peters 1996). HSA can spread to the extravascular space and return to the plasma. It should be noted that although albumin is a serum protein, it also locates extravascular, because it leaves the blood vessels through the pores of sinusoidal or fenestrated endothelium and goes to the liver, pancreas, small intestine, skin, muscle, and bone marrow. Albumin is degraded more quickly if it is denatured or altered – damaged structure of HSA is a signal to removal albumin from the circulation (Peters 1996). It was found that in normal condition, the kidney degrades large amounts of albumin and its fragments can appear in the urine and renal loss of albumin is about 6% of all synthesized HSA (Gudehithlu et al. 2004). Albumin excretion in the urine is correctly if it is less than 20 mg of albumin per day (Peterson et al. 1969). About 10% of systemic albumin loss or clearance occurs in gastrointestinal tract, but it may be intensified in various disease states including Crohn's disease, celiac disease, sarcoidosis, heart failure and portal hypertension. Studies have shown that labelled albumin is trapped in the lysosomes of fibroblasts, which indicate that albumin hydrolysis mainly occurs in skin and muscle (Levitt & Levitt 2016). It is estimated that under normal conditions, in a healthy individual, albumin is secreted and degraded at a rate 14 g/day (Kragh-Hansen 2016).

The structure of HSA undergoes various modifications including glycation, cysteinylation, S-nitrosylation, S-guanylation which alter structure and binding properties of HSA. Glycation is a nonenzymatic reaction between a ketone or aldehyde sugar group, mainly glucose, with amino group of a protein, resulting in formation of stable advanced glycation end products. The predominant site of glycation of HSA in native conformation is lysine-525 and it is responsible for one-third of overall glycation (Iberg & Flückiger, 1986) and among the total of 59 lysine residues, 34 may be glycated. In a healthy person, the ratio of glycated HSA is about 1-10% compared to no glycated serum albumin (Shaklai et al. 1984), but this increases in uncontrolled diabetes mellitus up to 90% (Kisugi et al. 2007). Measurement of glycated HSA reflects short-term glycemia and it is used as an additional marker, next to glycated hemoglobin, to monitor and screen of diabetic patients (Freitas et al. 2017. Glycation alters the structure and function of HSA including molecular mass increase, resistance to proteolytic enzymes, affinity for many drugs, and induction of protein aggregation (Anguizola et al. 2013). Glycation of HSA induces changes of secondary and tertiary structure, which lead to a decrease in the of α-helix content and generation of amyloid-like aggregates with high cytotoxicity and ability to induce oxidative stress and apoptosis (Iannuzzi et al. 2014).

Human serum albumin in blood consists of two forms depending on redox state – mainly reduced (mercaptalbumin) and oxidized (nonmercaptalbumin) HSA. The former comprises about 70% of total HSA and it has 17 disulphide bonds and one free thiol group at cysteine at position 34. On the other hand, the oxidized form of HSA has various modifications at cysteine-34 and occurs in two subtype forms– reversible and irreversible. Cysteine-34 undergoes oxidation which leads to changes in the function of HAS, for example increased protease susceptibility, decreased affinity for binding ligands and antioxidant activity (Kawakami et al. 2006). It should be noted that some of these oxidative modifications are reversible, but only 2-5% of HSA exists in irreversible oxidized form. Notwithstanding, the oxidized form of albumin prevails in many pathological states with severe oxidative stress as well as with aging. A minor fraction of HSA occurs in irreversible form of oxidized HSA where thiol groups are modified to more oxidized forms like sulfidic and sulfonic acids. Initially, oxidation causes formation of sulfenic acid from reaction between thiol group and oxidizing agents, such as hydrogen peroxide and peroxinitrite. Then, sulfenic acid may interact with thiol group or an amine group in reversible reaction or may be oxidized in irreversible reaction to more stable sulfidic and sulfonic acids (Turell et al. 2009). However, after reaction of sulfenic acid with low molecular-mass-thiol, such as glutathione, oxidized albumin may return to reduced form of HSA. From this it follows that HSA plays a protective role against oxidative stress by regulating glutathione level. Radicals of nitrosative stress, for instance, nitric oxide, cause a conversion of the reduced form of HSA to nitroso-HSA form; however, it can be converted back to mercaptalbumin indicating that albumin is a NO reservoir (Quinlan et al. 2005). Notably, about 80% of blood nitric oxide is transported in form of S-nitrosothiol bound by cysteine-34, but also tryptophan and tyrosine residues could bind this molecule (Peters 1996).

Summarizing, cysteine-34 acts as scavenger of multiple oxygen and nitrogen free radicals, such as hydrogen peroxide, superoxide, peroxinitrite, and hypochlorous acid, and with methionine residues are responsible for about 40-80% of antioxidant activity of albumin (Rocheet al. 2008; Wang et al. 2015).

Cysteinylation, the major mechanism of protein–thiol modification of HSA, mainly of cysteine-34, causes an increase in surface hydrophobicity and a decrease in affinity for ligand binding, especially in Sudlow's site II. Likewise, non-enzymatic N-homocysteinylation affects lysine residues – lysine-4, lysine-12 and lysine-525 (Fanali et al. 2012). Cysteinylation of HSA leads to structural modifications in regions surrounding drug binding sites and subsequent oxidation of other amino acids residues, for instance, cause an increase in carbonyl protein content. What is more, measurement of cysteinylation of cysteine-34 may be a useful marker in monitoring chronic cirrhotic patients (Nagumo et al. 2014). Another modification of cysteine residue, especially cysteine-34 and cysteine-410, is S-nitrosylation which causes the increase in the affinity of HSA for copper and ability of HSA to interact with fatty acids (Ishima et al. 2007b). In addition, S-nitrosylation weakens interaction between HSA and L-tryptophan, progesterone, ascorbate, zinc and iron (Ishima et al. 2007a). Cysteine also undergoes S-guanylation, a novel posttranslational modification of HSA, which is an interaction between 8-nitroguanosine 3′,5′-cyclic monophosphate group and sulfhydryl groups of protein. It is still unclear how this modification affects structure and affinity for ligand binding, but it may act as endogenous antibacterial agent (Ishima et al. 2012). Furthermore, albumin is susceptible to acetylation. When HSA interacts with aspirin, acetyl group binds covalently to amino group of lysine, what results in modulation of binding properties. This characteristic of HSA is not only structural but also functional because it accounts for enzymatic activities of plasma and different ligand-binding properties.

Human serum albumin displays enzymatic activity, for instance, it acts as esterase, thioesterase and thiol peroxidase, making it useful for medical therapeutic purposes.

Firstly, HSA has two reactive amino acid residues – tyrosine-411 and lysine-199, which are essential for the esterase activity of protein. However, depending on the involving site, the hydrolysis of esters to an acid and an alcohol proceeds differently. Hydrolysis using a water molecule takes place at position 411, while at lysine-199 a water molecule is not involved in this process. Moreover, both products of reaction are released in case of esterase activity involving tyrosine-411, whereas one product reacts covalently to lysine-199 and second one is released. Binding of product to lysine-199 and forming a stable adduct has its consequences, because lysine-199 is located at the entrance to Sudlow's site I, which may affect the ability to bind ligands (Peters 1996; Kragh-Hansen 2013). A study by Walker has shown that lysine-199 is involved in a split of acetylsalicylic acid to salicylic acid and acetyl group which binds to this amino acid residue and only this amino acid residue is mediated in hydrolysis of aspirin (Walker 1976). Acetylated lysine-199 remains stable for at least 21 days. Reaction between HSA and aspirin also results in modification of multiple lysine residues – lysine-402, lysine-519 and lysine-545 (Liyasova et al. 2010).

The esterase activity of albumin is slow, but it accounts for 40% of the total plasma esterase activity and it shows this activity against aspirin, long and short-chain fatty acid esters, ketoprofen glucuronide, cyclophosphamide or nicotinic acid esters (Wang et al. 2015).

HSA has thioesters, glutathione peroxidase, and cysteine peroxidase activities which are responsible for the significant capability of HSA to remove reactive oxygen species e.g., hydrogen peroxide (Cha et al. 1996). Thioesters activity results from presence of cysteine-34 in subdomain IA and participates in degradation of disulfiram, which is an aldehyde dehydrogenase inhibitor used in case of alcohol addiction. Similar to the esterase activity of HSA involving lysine-199, this drug is split into two molecules, one of which is released while second one is bound to cysteine-34 (Pedersen & Jacobsen, 1980). Recently, it has been discovered that HSA can hydrolyse RNA to nucleoside-3'- and nucleoside-2'-phosphates. It is supposed that the protein may participate in degradation of extracellular RNA, however efficiency of RNA hydrolysis is lower than in case of ribonucleases (Gerasimova et al. 2010). HSA also displays detoxification activity. Due to cysteine, tryptophan, arginine and tyrosine residues, HSA displays carboxylase activity in hydrolysis of carbamate carbaryl pesticides (Sogorb et al. 2004). Likewise, albumin hydrolyses highly toxic organophosphorus chemical compounds - esters of phosphoric or phosphonic acids (Li et al. 2008), paraoxon, chlorpyrifos oxon, and diazoxon (Sogorb et al. 2008). In view of that HSA binds covalently with tyrosine-411 or tyrosine-150 one of the breakdown products of these compounds, albumin may be a useful sensitive marker of exposure to widely used pesticides (Tarhoni et al. 2008).

Human serum albumin has ability to reversible or irreversible interaction usually at one of two primary sites with ligands including various molecules of endogenous and exogenous origin, hence, it is characterized by its extraordinary ligand binding capacity. Binding to HSA allows to control the concentration of a drug, ensures the availability of medicines for a long duration of action.

On the other hand, binding of ligands causes alternation in the polarity and volume of drug binding sites. HSA also transports a lot of compounds e.g., metabolites, fat soluble hormones, drugs and essential metal ions in the circulatory system to their destinations in tissues. Albumin binds not only compounds with low molecular mass, but also those with high molecular weight i.e., peptides and protein. Many proteins, such as angiotensinogen, apolipoproteins, ceruloplasmin, hemoglobin, plasminogen, prothrombin and transferrin may be bound and transported by albumin.

HSA is involved in transition metal ions transport of copper, zinc, iron, nickel, cadmium, mercury and vanadium. Binding of copper by albumin also leads to limit redox activity and the generation of reactive oxygen radicals. The sulfhydryl group of cysteine-34 is also involved in binding gold and platinum as well as nitric oxide (Bal et al. 2013). It has been noted that in the HSA structure, there are three primary binding sites which matches for the different metal geometries. The first one is located at the N-terminus involving aspartic acid-1, alanine-2 and histidine-3, the second is formed by the free cysteine-34, while the third site is formed by histidine-67, asparagine-99, histidine-247 and aspartic acid-249 residues (Fanali et al. 2012).

Sudlow's site I bind various drugs, such as salicylate, warfarin, digitoxin, indomethacin, tolbutamide, oxacillin, and benzylpenicillin, while diazepam, ibuprofen, naproxen and clofibrate are bound to Sudlow's site II (Peters 1996). Due to the opportunity to reduce side effects of treatment and improve the delivery of drugs, human serum albumin is commonly used as a medication carrier.

Fatty acids are bound to the two principal binding sites, but they may react diversely depending on the type of binding compounds. While Sudlow's site I is likely to be a main site for fatty acids with medium-chain, Sudlow's site II has a high affinity for long-chain fatty acids. Because there are additional binding sites, there are a total of 11 binding sites for medium and long-chain fatty acids for instance in sub-domain IB or between sub-domains IA and IIA (Bhattacharya et al. 2000). Fatty acids have high affinity for sub-domain IIIA, therefore in case of presence of them, other ligands are bound to sub-domains IB and IIA. On the other hand, fatty acids have low affinity for sub-domain IIA, so drugs successfully take their binding sites (Simard et al. 2006).

Binding of fatty acid to HSA causes a change in conformation of the protein and affects the binding of other ligands for example thyroxine and other hydrophobic compounds. The thyroid hormone mainly binds to Sudlow's site I, however it is known that in HSA structure are four binding sites for thyroxine located in subdomains IIA, IIIA, and IIIB. Hormone and fatty acids compete at all binding sites, although in case of binding of fatty acids thyroxine binds to additional fifth binding site between domains I and III (Petitpas et al. 2003). Furthermore, one molecule of HSA can carry toxic metabolites e.g., bilirubin, in which at least two molecules are bound in sub-domain IB and transported to the hepatocytes (Jacobsen & Brodersen 1983). Albumin also interacts with phenol red in bilirubin binding site, which can be a useful marker to study drugs binding sites in patient sera solutions with abnormal molar ratio between HSA and bilirubin (Sochacka 2015). Acetylation of HSA by aspirin increases affinity of HSA for phenylbutazone, inhibits bilirubin binding and decreases affinity for prostaglandin. Acetylated albumin also shows lower esterase activity (Liyasova et al. 2010). Likewise, glycation also alters binding properties of HSA. Glycation of lysine at position of 199 causes an increase in warfarin binding, while it decreases affinity of HSA to bilirubin (Fanali et al. 2012).

Similarly, at physiological pH, HSA binds other dyes such as Congo Red, what may have impact on the function of albumin and the elimination of toxic dye from the body after exposure (Patel and Kerman 2018).

Alinovskaya et al. has demonstrated for the first time that HSA has two sites to which DNA or RNA is bound. Moreover, HSA shows specific properties for nucleic acid recognition (Alinovskaya et al. 2018). Likewise, HSA interacts with folic acid, leading to a partial unfolding of the structure of protein and the reduction of lifetime of HSA (Chilom et al. 2018).

In summary, human serum albumin, being the most abundant protein of plasma, performs many important functions. HSA structure may undergo several modifications that could affect its primary properties. The protein displays enzymatic activity contributing to the conversion of many chemical compounds including medicines. Moreover, it shows antioxidant activity and reacts with free oxygen and nitrogen radicals, such as hydrogen peroxide or peroxynitrite. Human serum albumin also plays an extraordinary role in the binding and transport of endogenous and exogenous ligands which causes more and more interest of HSA as drug carrier in medicine.

ACKNOWLEDGMENTS

Dorota Bartusik-Aebisher acknowledges support from the National Center of Science NCN (New drug delivery systems-MRI study, Grant OPUS-13 number 2017/25/B/ST4/02481).

REFERENCES

Akirov, A., Masri-Iraqi, H., Atamna, A., Shimon, I. (2017). Low albumin levels are associated with mortality risk in hospitalized patients. *American Journal of Medicine,* 130(12): 1465. e11-e19.

Alinovskaya, L. I., Sedykh, S. E., Ivanisenko, N. V., Soboleva, S. E., Nevinsky, G. A. (2018). How human serum albumin recognizes DNA and RNA. *Journal of Biological Chemistry,* 399(4), 347-60.

Anguizola, J., Matsuda, R., Barnaby, O. S., Hoy, K. S., Wa, C., DeBolt, E., Koke, M., Hage, D. S. (2013). Review: Glycation of human serum albumin. *Clinica Chimica Acta,* 425, 64-76.

Bal, W., Sokołowska, M., Kurowska, E., Faller, P. (2013). Binding of transition metal ions to albumin: sites, affinities and rates. *Biochimica et Biophysica Acta,* 1830(12), 5444-55.

Bhattacharya, A. A., Grüne, T., Curry, S. (2000). Crystallographic analysis reveals common modes of binding of medium and long-chain fatty acids to human serum albumin. *Journal of Molecular Biology,* 303(5), 721-32.

Bramanti, E., Benedetti, E. (1996). Determination of the secondary structure of isomeric forms of human serum albumin by a particular frequency deconvolution procedure applied to Fourier transform IR analysis. *Biopolymers,* 38(5), 639-53.

Caraceni, P., Tufoni, M., Bonavita, M. E. (2013). Clinical use of albumin. *Blood Transfusion,* 11 Suppl 4, 18-25.

Carter, D. C., He, X. M., Munson, S. H., Twigg, P. D., Gernert, K. M., Broom, M. B., Miller, T. Y. (1989). Three-dimensional structure of human serum albumin. *Science,* 244(4909), 1195-8.

Cha, M. K., Kim, I. H. (1996). Glutathione-linked thiol peroxidase activity of human serum albumin: A possible antioxidant role of serum albumin in blood plasma. *Biochemical and Biophysical Research Communications,* 222, 619–25.

Chilom, C. G., Bacalum, M., Stanescu, M. M., Florescu, M. (2018). Insight into the interaction of human serum albumin with folic acid: A biophysical study. *Spectrochimica Acta Part A: Molecular and Biomolecular Spectroscopy,* 204, 648-56.

Choi, S., Choi, E. Y., Kim, D. J., Kim, J. H., Kim, T. S., Oh, S. W. (2005). A rapid, simple measurement of human albumin in whole blood using a fluorescence immunoassay (I). *Clin Chim Acta,* 339(1-2), 147-56.

Erickson, H.P. (2009). Size and shape of protein molecules at the nanometer level determined by sedimentation, gel filtration, and electron microscopy. *Biological Procedures Online,* 11, 32-51.

Fagan, T. C., Oexmann, M. J. (1987). Effects of high protein, high carbohydrate, and high fat diets on laboratory parameters. *The Journal of the American College of Nutrition*, 6(4), 333-43.

Fanali, G., di Masi, A., Trezza, V., Marino, M., Fasano, M., Ascenzi, P. (2012). Human serum albumin: from bench to bedside. *Molecular Aspects of Medicine*, 33(3), 209-90.

Fasano, M., Curry, S., Terreno, E., Galliano, M., Fanali, G., Narciso, P., Notari, S., Ascenzi, P. (2005). The extraordinary ligand binding properties of human serum albumin. *IUBMB Life*, 57(12), 787-96.

Feldman, M., Soni, N., Dickson, B. (2005). Influence of hypoalbuminemia or hyperalbuminemia on the serum anion gap. *Journal of Laboratory and Clinical Medicine*, 146(6), 317-20.

Freitas, P. A. C., Ehlert, L. R., Camargo, J. L. (2017). Glycated albumin: a potential biomarker in diabetes. *Archives of Endocrinology and Metabolism* 61(3), 296-304.

Gatta, A., Verardo, A., Bolognesi, M. (2012). Hypoalbuminemia. *Internal and Emergency Medicine*, *7 Suppl* 3, 193-9.

Gerasimova, Y. V., Bobik, T. V., Ponomarenko, N. A., Shakirov, M. M., Zenkova, M. A., Tamkovich, N. V., Popova, T. V., Knorre, D. G., Godovikova, T. S. (2010). RNA-hydrolyzing activity of human serum albumin and its recombinant analogue. *Bioorganic & Medicinal Chemistry Letters*, 20(4), 1427-31.

Glaumann, H., Ericsson, J. L. (1970). Evidence for the participation of the Golgi apparatus in the intracellular transport of nascent albumin in the liver cell. *Journal of Cell Biology*, 47(3), 555-67.

Gudehithlu, K. P., Pegoraro, A. A., Dunea, G., Arruda, J. A., Singh, A. K. (2004). Degradation of albumin by the renal proximal tubule cells and the subsequent fate of its fragments. *Kidney International*, 65(6), 2113-22.

Hawkins, J. W., Dugaiczyk, A. (1982). The human serum albumin gene: structure of a unique locus. *Gene*, 19(1), 55-8.

Iannuzzi, C., Irace, G., Sirangelo, I. (2014). Differential effects of glycation on protein aggregation and amyloid formation. *Frontiers in Molecular Biosciences*, 1, 9.

Iberg, N., Flückiger, R. (1986). Nonenzymatic glycosylation of albumin in vivo. Identification of multiple glycosylated sites. *Journal of Biological Chemistry,* 261, 13542-5.

Ishima, Y., Akaike, T., Kragh-Hansen, U., Hiroyama, S., Sawa, T., Maruyama, T., Kai, T., Otagiri, M. (2007a). Effects of endogenous ligands on the biological role of human serum albumin in S-nitrosylation. *Biochemical and Biophysical Research Communications,* 364(4), 790-5.

Ishima, Y., Hoshino, H., Shinagawa, T., Watanabe, K., Akaike, T., Sawa, T., Kragh-Hansen, U., Kai, T., Watanabe, H., Maruyama, T., Otagiri, M. (2012). S-guanylation of human serum albumin is a unique posttranslational modification and results in a novel class of antibacterial agents. *Journal of Pharmaceutical Sciences,* 101(9), 3222-9.

Ishima, Y., Sawa, T., Kragh-Hansen, U., Miyamoto, Y., Matsushita, S., Akaike, T. & Otagiri, M. (2007b). S-Nitrosylation of human variant albumin Liprizzi (R410C) confers potent antibacterial and cytoprotective properties. *Journal of Pharmacology and Experimental Therapeutics, 320*(3), 969-77.

Jacobsen, J., Brodersen, R. (1983). Albumin-bilirubin binding mechanism. *The Journal of Biological Chemistry,* 258(10), 6319-26.,

Jang, Y., Lee, S. T., Kim, T. J., Jun, J. S., Moon, J., Jung, K. H., Park, K. I., Chu, K., Lee, S. K. (2018). High albumin level is a predictor of favorable response to immunotherapy in autoimmune encephalitis. *Scientific Reports*, 8(1), 1012.

Kawakami, A., Kubota, K., Yamada, N., Tagami, U., Takehana, K., Sonaka, I., Suzuki, E., Hirayama, K. (2006). Identification and characterization of oxidized human serum albumin. A slight structural change impairs its ligand-binding and antioxidant functions. *FEBS Journal,* 273(14), 3346-57.

Kisugi, R., Kouzuma, T., Yamamoto, T., Akizuki, S., Miyamoto, H., Someya, Y., Yokoyama, J., Abe, I., Hirai, N., Ohnishi, A. (2007). Structural and glycation site changes of albumin in diabetic patient with very high glycated albumin *Clinica Chimica Acta,* 382(1-2), 59-64.

Kragh-Hansen, U. (2013). Molecular and practical aspects of the enzymatic properties of human serum albumin and of albumin-ligand complexes. *Biochimica et Biophysica Acta,* 1830(12), 5535-44.

Kragh-Hansen, U. (2016). Human serum albumin: a multifunctional protein. In: Otagiri M., Chuang V. (eds) *Albumin in Medicine.* Singapore: Springer.

Levitt, D. G., Levitt, M. D. (2016). Human serum albumin homeostasis: a new look at the roles of synthesis, catabolism, renal and gastrointestinal excretion, and the clinical value of serum albumin measurements. *International Journal of General Medicine,* 9, 229-55.

Li, B., Nachon, F., Froment, M. T., Verdier, L., Debouzy, J. C., Brasme, B., Gillon, E., Schopfer, L. M., Lockridge, O., Masson, P. (2007). Binding and hydrolysis of soman by human serum albumin. *Chemical Research in Toxicology,* 21(2), 421-31.

Liyasova, M. S., Schopfer, L. M., Lockridge, O. (2010). Reaction of human albumin with aspirin in vitro: mass spectrometric identification of acetylated lysines 199, 402, 519, and 545. *Biochemical Pharmacology,* 79(5), 784-91.

Meloun, B., Morávek, L., Kostka, V. (1975). Complete amino acid sequence of human serum albumin. *FEBS Letters* 58(1), 134-7.

Minghetti, P. P., Ruffner, D. E., Kuang, W. J., Dennison, O. E., Hawkins, J. W., Beattie, W. G., Dugaiczyk, A. (1986). Molecular structure of the human albumin gene is revealed by nucleotide sequence within q11-22 of chromosome 4. *Journal of Biological Chemistry,* 261(15), 6747-57.

Nagumo, K., Tanaka, M., Chuang, V. T., Setoyama, H., Watanabe, H., Yamada, N., Kubota, K., Tanaka, M., Matsushita, K., Yoshida, A., Jinnouchi, H., Anraku, M., Kadowaki, D., Ishima, Y., Sasaki, Y., Otagiri, M., Maruyama, T. (2014). Cys34-cysteinylated human serum albumin is a sensitive plasma marker in oxidative stress-related chronic diseases. *PLoS One,* 9(1), e85216.

Patel, B. R., Kerman, K. (2018). Calorimetric and spectroscopic detection of the interaction between a diazo dye and human serum albumin. *Analyst,* 143(16), 3890-9.

Pedersen, A. O., Jacobsen, J. (1980). Reactivity of the thiol group in human and bovine albumin at pH 3–9, as measured by exchange with 2,20-dithiodipyridine. *European Journal of Biochemistry,* 106, 291-5.

Peters, T. (1996). *All about albumin. Biochemistry, genetics and medical applications.* London: Academic Press.

Peterson, P. A., Evrin, P. E. & Berggård, I. (1969). Differentiation of glomerular, tubular, and normal proteinuria: determinations of urinary excretion of beta-2-macroglobulin, albumin, and total protein. *Journal of Clinical Investigation,* 48(7), 1189-98.

Petitpas, I., Petersen, C. E., Ha, C. E., Bhattacharya, A. A., Zunszain, P. A., Ghuman, J., Bhagavan, N. V. & Curry, S. (2003). Structural basis of albumin-thyroxine interactions and familial dysalbuminemic hyperthyroxinemia. *Proceedings of the National Academy of Sciences of the USA,* 100(11), 6440-5.

Quinlan, G. J., Martin, G. S. & Evans, T. W. (2005). Albumin: biochemical properties and therapeutic potential. *Hepatology,* 41, 1211-9.

Roche, M., Rondeau, P., Singh, N. R., Tarnus, E. & Bourdon, E. (2008). The antioxidant properties of serum albumin. *FEBS Letters,* 582, 1783-7.

Shaklai, N., Garlick, R. L. & Bunn, H. F. (1984). Nonenzymatic glycolsylation of human serum albumin alters its conformation and function. *Journal of Biological Chemistry,* 259(6), 3812-7.

Simard, J. R., Zunszain, P. A., Hamilton, J. A. & Curry, S. (2006). Location of high and low affinity fatty acid binding sites on human serum albumin revealed by NMR drug-competition analysis. *Journal of Molecular Biology,* 361, 336–51.

Sochacka, J. (2015). Application of phenol red as a marker ligand for bilirubin binding site at subdomain IIA on human serum albumin. *Journal of Photochemistry and Photobiology B: Biology,* 151, 89-99.

Sogorb, M. A., Carrera, V. & Vilanova, E. (2004). Hydrolysis of carbaryl by human serum albumin. *Archives of Toxicology,* 78(11), 629-34.

Sogorb, M. A., García-Argüelles, S., Carrera, V. & Vilanova, E. Serum albumin is as efficient as paraxonase in the detoxication of paraoxon at toxicologically relevant concentrations. *Chemical Research in Toxicology,* 21(8), 1524-9.

Sugio, S., Kashima, A., Mochizuki, S., Noda, M. & Kobayashi, K. (1999). Crystal structure of human serum albumin at 2.5 A resolution. *Protein engineering, design & selection,* 12(6), 439-46.

Tarhoni, M. H., Lister, T., Ray, D. E. & Carter, W. G. Albumin binding as a potential biomarker of exposure to moderately low levels of organophosphorus pesticides. *Biomarkers,* 13(4), 343-63.

Turell, L., Carballal, S., Botti, H., Radi, R. & Alvarez, B. (2009). Oxidation of the albumin thiol to sulfenic acid and its implications in the intravascular compartment. *Brazilian Journal of Medical and Biological Research,* 42(4), 305-11.

Walker, J. E. (1976). Lysine residue 199 of human serum albumin is modified by acetylsalicylic acid. *FEBS Letters,* 66, 173-5.

Wang, Y., Wang, S. & Huang, M. (2015). Structure and enzymatic activities of human serum albumin. *Current Pharmaceutical Design,* 21(14), 1831-6.

In: Human Serum Albumin
Editor: Dianne Cohen

ISBN: 978-1-53614-787-2
© 2019 Nova Science Publishers, Inc.

Chapter 2

HUMAN SERUM ALBUMIN AND ANTIOXIDANT PROPERTIES

David Aebisher[*], *Dorota Bartusik-Aebisher and Łukasz Ożóg*
Faculty of Medicine, University of Rzeszów, Rzeszów, Poland

ABSTRACT

Human Serum Albumin represents an important circulating antioxidant. This review brings together major recent insights on Human Serum Albumin antioxidant properties. Here we describe the role of albumin in free radical trapping activities and as an oxyradical scavenger. A discussion of recent advances of the use of Human Serum Albumin antioxidant properties to make drugs detectable in vivo is also presented.

Keywords: antioxidant properties, Human Serum Albumin, oxyradical scavenger

---

[*] Corresponding Author Email: daebisher@ur.edu.pl.

INTRODUCTIONS TO FUNCTIONS OF HUMAN SERUM ALBUMIN

The main roles of albumin in the human body is maintaining a constant oncotic pressure - that is, regulating the amount of water in the blood and preventing its passage from the plasma to the tissue fluid. Albumins transport a huge amount of different small molecules, starting with some hormones (thyroxine, triiodothyronine, cortisol), drugs (including antibiotics, barbiturates), fatty acids, lipids and bile dyes (bilirubin), vitamins. Albumins also play a role in the transport of, e.g., nitric oxide. In comparison to other proteins (haptoglobin, transferrin) they are non-specific but key transporters. In addition, cations of various metals such as calcium (Ca), sodium (Na), magnesium (Mg), zinc (Zn), and potassium (K) can also bind to albumin and move in the body in this form.

Endogenous substances that depend on albumin for binding and transport are:

- bilirubin,
- divalent cations,
- fatty acids,
- free radical species,
- fat-soluble vitamins,
- hormones.

Exogenous substances that depend on albumin for binding and transport are:

- Antibiotics
- Anticoagulants
- Anti-inflammatory drugs
- Anticonvulsants
- Cardiovascular and renal drugs

- Central nervous system drugs (e.g., amitriptyline, chlorpromazine, thiopental)
- Hypoglycemic agents
- Radiocontrast media

Human serum albumin control extracellular antioxidant functions (Sitar et al. 2013). Moreover, biomarkers of oxidative protein damage can increase (Sitar et al. 2013; Armstrong et al. 1998).

The main functions of Human serum albumin are:

- controls the plasma oncotic pressure (Ascenzi et al. 2015)
- modulates fluid distribution between the body compartments (Ascenzi et al. 2015)
- represents the depot and carrier of endogenous and exogenous compounds (Ascenzi et al. 2015)
- increases the apparent solubility and lifetime of hydrophobic compounds (Ascenzi et al. 2015)
- affects pharmacokinetics of many drugs,
- inactivates toxic compounds (Ascenzi et al. 2015)
- induces chemical modifications of some ligands (Ascenzi et al. 2015)
- displays antioxidant properties (Ascenzi et al. 2015)
- shows enzymatic properties (Ascenzi et al. 2015)

Table 1 presents review on Human serum albumin and antioxidant properties.

Research done by Medina-Navarro and coworkers discussed the response of Human serum albumin to stress (Medina-Navarro et al. 2014). The generation of semialdehydes in Human serum albumin after treatment with methylglyoxal and glyoxal was reported by Arcanio and coworkers (Arcanjo et al. 2018; Kawai et al. 2018). Glycation and oxidative damage cause protein modifications, frequently observed in numerous diseases

(Roche et al. 2008). Human serum albumin has been used for a long time as a resuscitation fluid in critically ill patients (Taverna et al. 2013).

Table 1. Human Serum Albumin and antioxidant properties

Reference	Aim of the study
Armstrong et al. 1998	Human serum albumin reactive oxygen species scavenging potential and effects on total oxidative stress to spermatozoa
Colombo et al. 2012	review of studies about oxidized albumin and albumin redox state as a global biomarker for the redox state in the body in various diseases
Kaneko et al. 2012	Human serum albumin level and antioxidant potentials in idiopathic nephrotic syndrome
Iwao et al. 2012	quantitative evaluating of the role of cysteine and methionine residues in the antioxidant activity of Human Serum Albumin using recombinant mutants
Taverna et al. 2013	review of the mechanisms responsible for the specific antioxidant properties of Human Serum Albumin in different forms
Baraka-Vidot et al. 2013	comparing the antioxidant properties of albumin purified from diabetic patients to in vitro models of glycated albumin
Du et al. 2013	investigating of the influence of glucose on the interaction between flavones and Human serum albumin and the effect of glucose on the antioxidant potential of a flavone- Human serum albumin system
Bruschi et al. 2013	reviewing studies about the role of albumin in plasma antioxidant activity mainly utilizing in vitro models of oxidation
Anraku et al. 2013	review of chemical changes in Human Serum Albumin induced by oxidative damage and their relevance to human pathology
Kragh-Hansen et al. 2013	review of genetic variants and molecular aspects of Human serum albumin, functional consequences and potential uses
Abdulmalik et al. 2013	study of transferring C_{60} molecules from HP-β-CyD to Human serum albumin molecules, and an investigation of the reactive oxygen species on the behavior of the resulting C_{60}/ Human serum albumin
Li et al. 2013	study of the interaction between Human serum albumin and antioxidants: ascorbic acid, α-tocopherol, and proanthocyanidins
Medina-Navarro et al. 2014	the albumin redox state of human albumin isolated from diabetic patients with progressive renal damage
Ellidag et al. 2014	investigating oxidative stress and ischemia-modified albumin in chronic ischemic heart failure
Turell et al. 2014	developing an external pH gradient chromatofocusing procedure for analysis of the oxidation status of Human serum albumin in human plasma and biopharmaceutical products

Reference	Aim of the study
Toker et al. 2014	determining serum ischemia modified albumin level as marker of oxidative stress and serum superoxide dismutase activity as a marker of antioxidant defense in patients with fibromyalgia
Pavićević et al. 2014	a study of fatty acids binding to Human serum albumin and changes of reactivity and glycation level of Cysteine-34 free thiol group with methylglyoxal
Da Silveira et al. 2014	evaluating the concentrations of new biomarkers of oxidative stress, ischemia-modified albumin and ferric reducing ability of plasma
Gokara et al. 2014	a study of the binding mechanism of asiatic acid with Human serum albumin and its biological implications
Sayar et al. 2015	investigating ischemia modified albumin as an oxidative stress marker in children with celiac disease
Rosas-Díaz et al. 2015	examining changes in the antioxidant capacity of albumin and alterations of the albumin structural conformation in patients of diabetes nephropathy
Anraku et al. 2015	determining molecular information about the antioxidant properties of Human serum albumin
Li et al. 2015	a study of the ability of Human serum albumin to scavenge 2,2-diphenyl-1-picrylhydrazyl radical using UV-vis absorption spectra; investigating the interaction between Human serum albumin and DPPH in the absence and presence of eight popular antioxidants
Ascenzi et al. 2015	review of the heme-based catalytic properties of Human serum albumin and the structural bases of drug-dependent allosteric regulation
Lim et al. 2015	attempting to identify the albumin redox status in the serum of patients on peritoneal dialysis
Huang et al. 2016	screening the antioxidant capacity of albumins isolated from uremic patients or healthy volunteers
Arcanjo et al. 2018	molecular interactions between resveratrol and Human serum albumin; antioxidant and pro-oxidant actions on Human serum albumin in the presence of toxic diabetes metabolites

Research done by Baraka-Vidot and coworkers presents information on important albumin functions after glycation (Baraka -Vidot et al. 2013). Also, those studies emphasize the importance of in vivo model of glycation in studies relied to diabetes pathology (Baraka-Vidot et al. 2013). It is known that oxidation of Human serum albumin causes to oxidative modification of Human serum albumin (Huang et al. 2016; Sayar et al. 2015). Hemoglobin-Human serum albumin is an artificial O_2 carrier that can function as a red blood cell substitute. Hemoglobin- Human serum albumin cluster has antioxidant activities (Hosaka et al. 2014). Human serum albumin is an important carrier protein of many endogenous and exogenous molecules throughout the body. Recent reports suggest that the albumin redox state may serve as a global biomarker for the redox state in the body in various human diseases (Colombo et al. 2012). The influence of glucose on the interaction between flavones and HSA was investigated, as well as the effect of glucose on the antioxidant potential of a flavone-HSA system (Du et al. 2013). Ellidag and coworkers showed that oxidative stress may be a key factor in the development of hypoalbuminemia (Ellidag et al. 2014). Serum levels of biological antioxidant has correlated well with serum albumin levels (Kaneko et al. 2012). Literature data has provided evidence for the partial denaturation of albumin and exacerbated oxidative stress among patients in advanced stages of diabetes nephropathy before and even after dialysis (Rosas-Díaz et al. 2015). The function of proteins that act as antioxidants in biological systems subjected to oxidative stress in conditions such as inflammation and aging were discussed by Ankru et al. (Anraku et al. 2015). The pharmacokinetic activity of isovitexin in rat blood plasma and physicochemical forces that govern the interaction was reported (Caruso et al. 2014). Human serum albumin is main modulator of fluid distribution between body compartments (Al Bittar et al. 2014; Fanali et al. 2012). The interaction of Fe^{3+} and Cu^{2+} ions with Human Serum Albumin was assayed and showed that this increases the stability (Behbehani et al. 2014). The study showed that antioxidant property of Human serum albumin is increased as a result of its interaction with Cu^{2+} (Rezaei Behbehani et al. 2014). A new chlorogenate oxidovanadium complex was synthesized by using Schlenk methodology in the course of a reaction in which deprotonated chlorogenic acid ligand binds to oxidovanadium (IV) in a reaction experiment (Naso et al. 2014). HSA is associated with diabetic complications under glycemic range of diabetes mellitus (Raghav et al. 2016). Human serum albumin, the most abundant circulating protein in the plasma, exerts important antioxidant activities against oxidative damage which results in protein modification, and is observed in numerous diseases (Anraku et al. 2013). The ability of human serum albumin to scavenge radicals was investigated using UV-vis absorption spectra (Li et al. 2015). Both in vivo circulating Human serum albumin and pharmaceutical preparations are heterogeneous with respect to the oxidation state of Cys34 (Turell et al. 2014). Human serum albumin has been molecularly characterized at the protein and gene level. They can also give valuable molecular information about albumin binding sites, antioxidant and enzymatic properties, as well as stability (Kragh-Hansen et al. 2013). The aim of the study by Toker and coworkers was to determine serum ischemia modified albumin and malondialdehyde levels as markers of oxidative stress. The authors determine serum superoxide dismutase activity as a marker of antioxidant defense and their associations with clinical outcomes in patients with fibromyalgia (Toker et al. 2014). Fatty acids binding to human serum albumin lead to changes in Cys-34 thiol group accessibility and reactivity (Pavićević et al. 2014). The importance of cysteine and methionine residues for the antioxidant activity of Human serum albumin was investigated using recombinant Human serum albumin mutants, in which Cys34 and the six Met residues had been mutated to Ala (Iwao et al. 2012). The stability of solutions of fullerene C_{60} with human serum albumin (C_{60}/HSA) has been studied and reported on the preparation of stable C_{60}/HSA solutions that are formed via the formation of C_{60}/HP-β-CyD nanoparticles, i.e., by transferring C_{60} molecules from HP-β-CyD to human serum albumin molecules, and an investigation of the reactive oxygen species on the behavior of the resulting C_{60}/HAS (Abdulmalik et al. 2013). There is evidence in prostate cancer which indicates that inflammation and oxidative stress play a key role in the pathogenesis of this disease (Da Silveira et al. 2014). Albumin, the most abundant protein in the extracellular fluid, displays an important antioxidant activity. Increased levels of oxidized human serum albumin levels have been reported in the serum of patients with end-stage renal disease (Lim et al. 2015). The interaction between the three antioxidants ascorbic acid, α-tocopherol and proanthocyanidins and human serum albumin was reported (Li et al. 2013). Asiatic acid, a naturally occurring pentacyclictriterpenoid found in *Centella asiatica,* plays a major role in neuroprotection, anticancer, antioxidant, and hepatoprotective activities. Human serum albumin, a blood plasma protein, participates in the regulation of plasma osmotic pressure and transports endogenous and exogenous substances (Gokara et al. 2014). Human serum albumin contributes to the stabilization of (-)-epigallocatechin gallate in serum (Ishii et al. 2011). Human serum albumin binding profiles of the metal complexes were monitored using biophysical techniques including absorbance, fluorescence, circular dichromism and foster resonance energy transfer (Shakir et al. 2016). Human serum albumin -thioredoxin 1 fusion protein was designed to overcome the unfavorable pharmacokinetic and short pharmacological properties of thioredoxin 1(Tanaka et al. 2014; Anraku et al. 2008). Oxidative modifications of Human serum albumin alter its biological properties and may affect its antioxidant potential (Lim et al. 2013; Condezo-Hoyos et al. 2009). Both the preventive activity against the glutamine synthetase inactivation and peroxidase activity were completely abolished by the reactions of Human serum albumin with N-ethylmaleimide and iodoacetate, chemical modification agents for sulfhydryl of protein (Cha and Kim1996). Human serum albumin account for more than 40% of the antioxidant effect of Human serum albumin *in vivo* (Turell et al. 2013; Anraku et al. 2011). Complexes of Human serum albumin and poly(2-aminodisulfide-4-nitrophenol) increased the catalytic activity and act as potent protectors and activators of catalase during enzymatic degradation of H_2O_2 (Metelitsa and Eremin 2002). Human serum albumin possesses a free thiol group in Cys-34 that could be a site for hydropersulfide formation. Human serum albumin hydropersulfide of high purity as a positive control was prepared by treatment of Human serum albumin with Na_2S (Shibata et al. 2016; Bae et al. 2009). A combination of fluorescence spectroscopy, circular dichroism spectroscopy, and molecular modeling methods was used for recognition of the binding site (Goncharova et al. 2015; Anraku 2014; Krizková et al. 2006; Sakata et al. 2002). Antioxidant properties of human serum albumin may explain part of its beneficial role in various diseases related to free radical attack. In the present study, the antioxidant role of Cys and Met was studied by copper-mediated oxidation of human low-density lipoproteins and by free radical-induced blood hemolysis which essentially assessed metal-chelating and free radical scavenging activities (Bourdon et al. 2005). Lipophilic antioxidants of serum were extracted with n-hexane from an ethanolic solution of serum subjected to centrifugation (Huang et al. 2014; Apak et al. 2010; Yamada et al. 2008; Kawakami et al. 2006; Ihara et al. 2004; Soriani et al. 1994). Rate constants for nitroxide reduction by antioxidants in Human serum albumin were determined (Aspée et al. 2009; Kim et al. 2001; Bourdon et al. 1999; Popov and Lewin 1999). FT-IR and UV-visible spectroscopic methods were used to determine the ligand binding mode, the binding constant, and the effects of ligand complexation on protein secondary structure (Zhang et al. 2018; Arif et al. 2018; Matsushita et al. 2017; Kanakis et al. 2007; Filipe et al. 2004; Kon et al. 2004; Dobrian et al. 1993).

CONCLUSION

The oxidized products of some amino acids and proteins acquire antiradical properties Human serum albumin. An increased level of oxidized Human serum albumin may impair function in a number of pathological conditions.

ACKNOWLEDGMENTS

Dorota Bartusik-Aebisher acknowledges support from the National Center of Science NCN (New drug delivery systems-MRI study, Grant OPUS-13 number 2017/25/B/ST4/02481).

REFERENCES

Abdulmalik, A., Hibah, A., Zainy, B. M., Makoto, A., Daisuke, I., Masaki, O., Kaneto, U., Fumitoshi, H. (2013). Preparation of soluble stable C_{60}/human serum albumin nanoparticles via cyclodextrin complexation and their reactive oxygen production characteristics. *Life Sciences*, 93(7):277-82.

Al Bittar, S., Mora, N., Loonis, M., Dangles, O. (2014). Chemically synthesized glycosides of hydroxylated flavylium ions as suitable models of anthocyanins: binding to iron ions and human serum albumin, antioxidant activity in model gastric conditions. *Molecules: a Journal of Synthetic Chemistry and Natural Product Chemistry*, 19(12):20709-30.

Anraku, M., Shintomo, R., Taguchi, K., Kragh-Hansen, U., Kai, T., Maruyama, T., Otagiri, M. (2015). Amino acids of importance for the antioxidant activity of human serum albumin as revealed by recombinant mutants and genetic variants. *Life Sciences*, 134:36-41.

Anraku, M., Takeuchi, K., Watanabe, H., Kadowaki, D., Kitamura, K., Tomita, K., Kuniyasu, A., Suenaga, A., Maruyama, T., Otagiri, M. (2011). Quantitative analysis of cysteine-34 on the anitioxidative properties of human serum albumin in hemodialysis patients. *Journal of Pharmaceutical Sciences*, 100(9):3968-76.

Anraku, M., Kabashima, M., Namura, H., Maruyama, T., Otagiri, M., Gebicki, J. M., Furutani, N., Tomida, H. (2008). Antioxidant protection of human serum albumin by chitosan. *International Journal of Biological Macromolecules*, 43(2):159-64.

Anraku, M. (2014). Elucidation of the mechanism responsible for the oxidation of serum albumin and its application in treating oxidative stress-related diseases. *Yakugaku Zasshi: Journal of The Pharmaceutical Society of Japan*, 134(9):973-9.

Ascenzi, P., di Masi, A., Fanali, G., Fasano, M. (2015). Heme-based catalytic properties of human serum albumin. *Cell Death Discovery*, 1:15025.

Armstrong, J. S., Rajasekaran, M., Hellstrom, W. J., Sikka, S. C. (1998). Antioxidant potential of human serum albumin: role in the recovery of high quality human spermatozoa for assisted reproductive technology. *Journal of Andrology*, 19(4):412-9.

Apak, R., Güçlü, K., Ozyürek, M., Bektaşoğlu, B., Bener, M. (2010). Cupric ion reducing antioxidant capacity assay for antioxidants in human serum and for hydroxyl radical scavengers. *Methods in Molecular Biology*, 594:215-39.

Arcanjo, N. M. O., Luna, C., Madruga, M. S., Estévez, M. (2018). Antioxidant and pro-oxidant actions of resveratrol on human serum albumin in the presence of toxic diabetes metabolites: Glyoxal and methyl-glyoxal. *Biochimica et Biophysica Acta*, 1862(9):1938-47.

Arif, Z., Neelofar, K., Arfat, M. Y., Zaman, A., Tarannum, A., Parveen, I., Ahmad, S., Khan, M. A., Badar, A., Islam, S. N. (2018). Hyperglycemia induced reactive species trigger structural changes in human serum albumin of type 1 diabetic subjects. *International Journal of Biological Macromolecules*, 107(Pt B):2141-49.

Aspée, A., Orrego, A., Alarcón, E., López-Alarcón, C., Poblete, H., González-Nilo, D. (2009). Antioxidant reactivity toward nitroxide probes anchored into human serum albumin. A new model for studying antioxidant repairing capacity of protein radicals. *Bioorganic and Medicinal Chemistry Letters*, 19(22):6382-5.

Bae, M. J., Ishii, T., Minoda, K., Kawada, Y., Ichikawa, T., Mori, T., Kamihira, M., Nakayama, T. (2009). Albumin stabilizes (-)-epigallocatechin gallate in human serum: binding capacity and antioxidant property. *Molecular Nutrition and Food Research*, 53(6):709-15.

Baraka-Vidot, J., Guerin-Dubourg, A., Dubois, F., Payet, B., Bourdon, E., Rondeau, P. (2013). New insights into deleterious impacts of in vivo glycation on albumin antioxidant activities. *Biochimica et Biophysica Acta*, 1830(6):3532-41.

Bruschi, M., Candiano, G., Santucci, L., Ghiggeri, G. M. (2013). Oxidized albumin. The long way of a protein of uncertain function. *Biochimica et Biophysica Acta*, 1830(12):5473-9.

Bourdon, E., Loreau, N., Lagrost, L., Blache, D. (2005). Differential effects of cysteine and methionine residues in the antioxidant activity of human serum albumin. *Free Radical Research*, 39(1):15-20.

Bourdon, E., Loreau, N., Blache, D. (1999). Glucose and free radicals impair the antioxidant properties of serum albumin. *FASEB Journal: Official Publication of The Federation of American Societies for Experimental Biology*, 13(2):233-44.

Caruso, Í. P., Vilegas, W., de Souza, F. P., Fossey, M. A., Cornélio, M. L. (2014). Binding of antioxidant flavone isovitexin to human serum albumin investigated by experimental and computational assays. *Journal of Pharmaceutical and Biomedical Analysis*, 98:100-6.

Colombo, G., Clerici, M., Giustarini, D., Rossi, R., Milzani, A., Dalle-Donne, I. (2012). Redox albuminomics: oxidized albumin in human diseases. *Antioxidants and Redox Signaling*, 17(11):1515-27.

Condezo-Hoyos, L., Abderrahim, F., Conde, M. V., Susín, C., Díaz-Gil, J. J., González, M. C., Arribas, S. M. (2009). Antioxidant activity of liver growth factor, a bilirubin covalently bound to albumin. *Free Radical Biology and Medicine*, 46(5):656-62.

Cha, M. K. and Kim, I. H. (1996). Glutathione-linked thiol peroxidase activity of human serum albumin: a possible antioxidant role of serum albumin in blood plasma. *Biochemical and Biophysical Research Communications*, 222(2):619-25.

Da Silveira, R. A., Hermes, C. L., Almeida, T. C., Bochi, G. V., De Bona, K. S., Moretto, M. B., Moresco, R. N. (2014). Ischemia-modified albumin and inflammatory biomarkers in patients with prostate cancer. *Clinical Laboratory*, 60(10):1703-8.

Du, S., Xie, Y., Chen, X. (2013). Influence of glucose on the human serum albumin-flavone interaction and their antioxidant activity. *Molecular bioSystems*, 9(1):55-60.

Dobrian, A., Mora, R., Simionescu, M., Simionescu, N. (1993). In vitro formation of oxidatively-modified and reassembled human low-density lipoproteins: antioxidant effect of albumin. *Biochimica et Biophysica Acta*, 1169(1):12-24.

Ellidag, H. Y., Eren, E., Yılmaz, N., Cekin, Y. (2014). Oxidative stress and ischemia-modified albumin in chronic ischemic heart failure. *Redox Report: Communications in Free Radical Research*, 19(3):118-23.

Fanali, G., di Masi, A., Trezza, V., Marino, M., Fasano, M., Ascenzi, P. (2012). Human serum albumin: from bench to bedside. *Molecular Aspects of Medicine*, 33(3):209-90.

Filipe, P., Morlière, P., Patterson, L. K., Hug, G. L., Mazière, J. C., Freitas, J. P., Fernandes, A., Santus, R. (2004). Oxygen-copper (II) interplay in the repair of semi-oxidized urate by quercetin bound to human serum albumin. *Free Radical Research*, 38(3):295-301.

Gokara, M., Malavath, T., Kalangi, S. K., Reddana, P., Subramanyam, R. (2014). Unraveling the binding mechanism of Asiatic acid with human serum albumin and its biological implications. *Journal of Biomolecular Structure and Dynamics*, 32(8):1290-302.

Goncharova, I., Jašprová, J., Vítek, L., Urbanová, M. (2015). Photo-isomerization and oxidation of bilirubin in mammals is dependent on albumin binding. *Analytical Biochemistry*, 490:34-45.

Huang, C. Y., Liou, S. Y., Kuo, W. W., Wu, H. C., Chang, Y. L., Chen, T. S. (2016). Chemiluminescence analysis of antioxidant capacity for serum albumin isolated from healthy or uremic volunteers. *Luminescence: The Journal of Biological and Chemical Luminescence*, 31(8):1474-8.

Hosaka, H., Haruki, R., Yamada, K., Böttcher, C., Komatsu, T. (2014). Hemoglobin-albumin cluster incorporating a Pt nanoparticle: artificial O2 carrier with antioxidant activities. *Public Library of Science One PLoS 1*, 9(10):e110541.

Huang, Y., Shuai, Y., Li, H., Gao, Z. (2014). Tyrosine residues play an important role in heme detoxification by serum albumin. *Biochimica et Biophysica Acta*, 1840(3):970-6.

Ishii, T., Ichikawa, T., Minoda, K., Kusaka, K., Ito, S., Suzuki, Y., Akagawa, M., Mochizuki, K., Goda, T., Nakayama, T. (2011). Human serum albumin as an antioxidant in the oxidation of (-)-epigallocatechin gallate: participation of reversible covalent binding for interaction and stabilization. *Bioscience, Biotechnology, and Biochemistry*, 75(1):100-6.

Ihara, H., Hashizume, N., Hasegawa, T., Yoshida, M. (2004). Antioxidant capacities of ascorbic acid, uric acid, alpha-tocopherol, and bilirubin can be measured in the presence of another antioxidant, serum albumin. *Journal of Clinical Laboratory Analysis*, 18(1):45-9.

Iwao, Y., Ishima, Y., Yamada, J., Noguchi, T., Kragh-Hansen, U., Mera, K., Honda, D., Suenaga, A., Maruyama, T., Otagiri, M. (2012). Quantitative evaluation of the role of cysteine and methionine residues in the antioxidant activity of human serum albumin using recombinant mutants. *International Union of Biochemistry and Molecular Biology Life*, 64(5):450-4.

Kanakis, C. D., Tarantilis, P. A., Tajmir-Riahi, H. A., Polissiou, M. G. (2007). Crocetin, dimethylcrocetin, and safranal bind human serum albumin: stability and antioxidative properties. *Journal of Agricultural and Food Chemistry*, 55(3):970-7.

Kawai, Y., Masutani, K., Torisu, K., Katafuchi, R., Tanaka, S., Tsuchimoto, A., Mitsuiki, K., Tsuruya, K., Kitazono, T. (2018). Association between serum albumin level and incidence of end-stage renal disease in patients with Immunoglobulin A nephropathy: A possible role of albumin as an antioxidant agent. *Public Library of Science One PLoS 1*, 13(5):e0196655.

Kaneko, K., Kimata, T., Tsuji, S., Shimo, T., Takahashi, M., Tanaka, S. (2012). Serum albumin level accurately reflects antioxidant potentials in idiopathic nephrotic syndrome. *Clinical and Experimental Nephrology*, 16(3):411-4.

Kawakami, A., Kubota, K., Yamada, N., Tagami, U., Takehana, K., Sonaka, I., Suzuki, E., Hirayama, K. (2006). Identification and characterization of oxidized human serum albumin. A slight structural change impairs its ligand-binding and antioxidant functions. *Federation of European Biochemical Societies Journal*, 273(14):3346-57.

Kim, I. G., Park, S. Y., Oh, T. J. (2001). Dithiothreitol induces the sacrificial antioxidant property of human serum albumin in a metal-catalyzed oxidation and gamma-irradiation system. *Archives of Biochemistry and Biophysics*, 388(1):1-6.

Kon, T., Tanigawa, T., Hayamizu, K., Shen, M., Tsuji, T., Naito, Y., Yoshikawa, T. (2004). Singlet oxygen quenching activity of human serum. *Redox Report: Communications In Free Radical Research*, 9(6):325-30.

Kragh-Hansen, U., Minchiotti, L., Galliano, M., Peters, T. Jr. (2013). Human serum albumin isoforms: genetic and molecular aspects and functional consequences. *Biochimica et Biophysica Acta*, 1830(12): 5405-17.

Krizková, L., Zitnanová, I., Mislovicová, D., Masárová, J., Sasinková, V., Duracková, Z., Krajcovic, J. (2006). Antioxidant and antimutagenic activity of mannan neoglycoconjugates: mannan-human serum albumin and mannan-penicillin G acylase. *Mutation Research. Fundamental and Molecular Mechanisms of Mutagenesis*, 606(1-2):72-9.

Lee, H., Cha, M. K., Kim, I. H. (2000). Activation of thiol-dependent antioxidant activity of human serum albumin by alkaline pH is due to the B-like conformational change. *Archives of Biochemistry and Biophysics*, 380(2):309-18.

Li, X., Chen, D., Wang, G., Lu, Y. (2013). Study of interaction between human serum albumin and three antioxidants: ascorbic acid, α-tocopherol, and proanthocyanidins. *European Journal of Medicinal Chemistry*, 70:22-36.

Li, X., Chen, D., Wang, G., Lu, Y. (2015). Probing the interaction of human serum albumin with DPPH in the absence and presence of the eight antioxidants. *Spectrochimica Acta. Part A, Molecular and Biomolecular Spectroscopy*, 137:1144-52.

Lim, P. S., Chen, H. P., Chen, C. H., Wu, M. Y., Wu, C. Y., Wu, T. K. (2015). Association between redox status of serum albumin and peritoneal membrane transport properties in patients on peritoneal dialysis. *Blood Purification*, 40(3):243-9.

Lim, P. S., Jeng, Y., Wu, M. Y., Pai, M. A., Wu, T. K., Liu, C. S., Chen, C. H., Kuo, Y. C., Chien, S. W., Chen, H. P. (2013). Serum oxidized albumin and cardiovascular mortality in normoalbuminemic hemodialysis patients: a cohort study. *Public Library of Science One PLoS 1*, 8(7):e70822.

Matsushita, S., Nishi, K., Iwao, Y., Ishima, Y., Watanabe, H., Taguchi, K., Yamasaki, K., Maruyama, T., Otagiri, M. (2017). Recombinant human serum albumin containing 3 copies of Domain I, has significant in vitro antioxidative capacity compared to the wild-type. *Biological and Pharmaceutical Bulletin*, 40(10):1813-17.

Medina-Navarro, R., Corona-Candelas, I., Barajas-González, S., Díaz-Flores, M., Durán-Reyes, G. (2014). Albumin antioxidant response to stress in diabetic nephropathy progression. *Public Library of Science One PLoS 1*, 9(9):e106490.

Metelitsa, D. I. and Eremin, A. N. (2002). Changes in the antioxidant properties of substituted phenol polydisulfides during interaction with human serum albumin. *Prikladnaia Biokhimiia i Mikrobiologiia*, 38(3):312-21.

Naso, L. G., Valcarcel, M., Roura-Ferrer, M., Kortazar, D., Salado, C., Lezama, L., Rojo, T., González-Baró, A. C., Williams, P. A., Ferrer, E. G. (2014). Promising antioxidant and anticancer (human breast cancer) oxidovanadium(IV) complex of chlorogenic acid. Synthesis, characterization and spectroscopic examination on the transport mechanism with bovine serum albumin. *Journal of Inorganic Biochemistry*, 135:86-99.

Pavićević, I. D., Jovanović, V. B., Takić, M. M., Penezić, A. Z., Aćimović, J. M., Mandić, L. M. (2014). Fatty acids binding to human serum albumin: Changes of reactivity and glycation level of Cysteine-34 free thiol group with methylglyoxal. *Chemico-Biological Interactions*, 224:42-50.

Popov, I. and Lewin, G. (1999). Photochemiluminescent detection of antiradical activity. VI. Antioxidant characteristics of human blood plasma, low density lipoprotein, serum albumin and amino acids during in vitro oxidation. *Luminescence: The Journal of Biological and Chemical Luminescence*, 14(3):169-74.

Raghav, A., Ahmad, J., Alam, K. (2016). Impact of glycation on structural and antioxidant function of human serum albumin: Relevance in diabetic complications. *Diabetes and Metabolic Syndrome*, 10(2):96-101.

Rezaei Behbehani, G., Gonbadi, K., Eslami, N. (2014). The effect of cupric and ferric ions on antioxidant properties of human serum albumi. *General Physiology and Biophysics*, 33(4):453-6.

Roche, M., Rondeau, P., Singh, N. R., Tarnus, E., Bourdon, E. (2008). The antioxidant properties of serum albumin. *Federation of European Biochemical Societies Letters*, 582(13):1783-7.

Rosas-Díaz, M., Camarillo-Cadena, M., Hernández-Arana, A., Ramón-Gallegos, E., Medina-Navarro, R. (2015). Antioxidant capacity and structural changes of human serum albumin from patients in advanced stages of diabetic nephropathy and the effect of the dialysis. *Molecular and Cellular Biochemistry*, 404(1-2):193-201.

Sayar, E., Özdem, S., Uzun, G., İşlek, A., Yılmaz, A., Artan, R. (2015). Total oxidant status, total antioxidant capacity and ischemia modified albumin levels in children with celiac disease. *The Turkish Journal of Pediatrics*, 57(5):498-503.

Sitar, M. E., Aydin, S., Cakatay, U. (2013). Human serum albumin and its relation with oxidative stress. *Clinical Laboratory*, 59(9-10):945-52.

Tanaka, R., Ishima, Y., Maeda, H., Kodama, A., Nagao, S., Watanabe, H., Chuang, V. T., Otagiri, M., Maruyama, T. (2014). Albumin fusion prolongs the antioxidant and anti-inflammatory activities of thioredoxin in mice with acetaminophen-induced hepatitis. *Molecular Pharmaceutics*, 11(4):1228-38.

Taverna, M., Marie, A. L., Mira, J. P., Guidet, B. (2013). Specific antioxidant properties of human serum albumin. *Annals of Intensive Care*, 3(1):4.

Turell, L., Botti, H., Bonilla, L., Torres, M. J., Schopfer, F., Freeman, B. A., Armas, L., Ricciardi, A., Alvarez, B., Radi, R. (2014). HPLC separation of human serum albumin isoforms based on their isoelectric points. *Journal of Chromatography. B, Analytical Technologies in the Biomedical and Life Sciences*, 944:144-151.

Toker, A., Kucuksen, S., Kucuk, A., Cicekler, H. (2014). Serum ischemia-modified albumin and malondialdehyde levels and superoxide dismutase activity in patients with fibromyalgia. *Clinical Laboratory*, 60(10):1609-15.

Turell, L., Radi, R., Alvarez, B. (2013). The thiol pool in human plasma: the central contribution of albumin to redox processes. *Free Radical Biology and Medicine*, 65:244-53.

Sakata, N., Moh, A., Takebayashi, S. (2002). Contribution of superoxide to reduced antioxidant activity of glycoxidative serum albumin. *Heart and Vessels*, 17(1):22-9.

Shakir, M., Hanif, S., Alam, M. F., Younus, H. (2016). Molecular hybridization approach of bio-potent Cu^{II}/Zn^{II} complexes derived from N, O donor bidentate imine scaffolds: Synthesis, spectral, human serum albumin binding, antioxidant and antibacterial studies. *Journal of Photochemistry and Photobiology. B, Biology*, 165:96-114.

Shibata, A., Ishima, Y., Ikeda, M., Sato, H., Imafuku, T., Chuang, V. T. G., Ouchi, Y., Abe, T., Watanabe, H., Ishida, T., Otagiri, M., Maruyama, T. (2016). Human serum albumin hydropersulfide is a potent reactive oxygen species scavenger in oxidative stress conditions such as chronic kidney disease. *Biochemical and Biophysical Research Communications*, 479(3):578-83.

Soriani, M., Pietraforte, D., Minetti, M. (1994). Antioxidant potential of anaerobic human plasma: role of serum albumin and thiols as scavengers of carbon radicals. *Archives of Biochemistry and Biophysics*, 312(1):180-8.

Yamada, N., Nakayama, A., Kubota, K., Kawakami, A., Suzuki, E. (2008). Structure and function changes of oxidized human serum albumin: physiological significance of the biomarker and importance of sampling conditions for accurate measurement. *Rinsho Byori. The Japanese Journal of Clinical Pathology*, 56(5):409-15.

Zhang, Y., Wu, S., Qin, Y., Liu, J., Liu, J., Wang, Q., Ren, F., Zhang, H. (2018). Interaction of phenolic acids and their derivatives with human serum albumin: Structure-affinity relationships and effects on antioxidant activity. *Food Chemistry*, 240:1072-1080.

In: Human Serum Albumin
Editor: Dianne Cohen

ISBN: 978-1-53614-787-2
© 2019 Nova Science Publishers, Inc.

Chapter 3

EVALUATION OF PHOTODYNAMIC AGENT ACTIVITY USING HUMAN SERUM ALBUMIN

Kazutaka Hirakawa[1,2,*]

[1]Applied Chemistry and Biochemical Engineering Course,
Department of Engineering,
Graduate School of Integrated Science and Technology,
Shizuoka University, Johoku, Naka-ku, Hamamatsu, Shizuoka, Japan
[2]Department of Optoelectronics and Nanostructure Science,
Graduate School of Science and Technology, Shizuoka University,
Johoku, Naka-ku, Hamamatsu, Shizuoka, Japan

ABSTRACT

Human serum albumin (HSA) is a water-soluble protein that can hydrophobically interact with organic molecules. HSA has one tryptophan residue and shows a characteristic fluorescence at around 350 nm under ultraviolet irradiation. Because tryptophan is easily oxidized by reactive oxygen species and/or photoexcited molecules through electron transfer, leading to fluorescence diminishment, a fluorometry of the HSA tryptophan residue is a useful tool to evaluate HSA oxidation. In light of

---

[*] Corresponding Author Email: hirakawa.kazutaka@shizuoka.ac.jp.

these HSA characteristics, we examined the photosensitizing activity of organic photosensitizers, including porphyrins and phenothiazine dyes. The purpose of this study using HSA is the development of agents (photosensitizers) for photodynamic therapy (PDT), which is a less invasive treatment of cancer and infections. Administered photosensitizers undergo visible light irradiation, resulting in the photochemical oxidation of targeting biomacromolecules to kill cancer cells and pathogenic bacteria. Therefore, photosensitized protein-damaging activity is an important property of photodynamic agents. Absorption spectral change can be used to examine the interaction between HSA and photosensitizers. Analysis of the relationship between the absorbance of photosensitizers in a visible light region and the concentration of HSA gives us the association constant and binding ratio. Fluorescence lifetime measurement of the tryptophan residue of HSA is a simple method to estimate the binding distance with photosensitizing molecules. After visible light irradiation of the sample solution consisting of HSA and photosensitizers, damaged contents of HSA can be estimated by measuring the intrinsic fluorescence intensity of the tryptophan residue. Using this simple system, the activity of photosensitizers and the damaging mechanisms can be elucidated. In this chapter, recent examples of the applications of HSA to study PDT photosensitizers are introduced.

Keywords: protein damage, binding interaction, tryptophan oxidation, singlet oxygen, electron transfer

1. INTRODUCTION

Human serum albumin (HSA) is a commercially available water-soluble protein. Its structure has been clarified by X-ray crystal structure analysis [1]. HSA consists of 585 amino acid residues (Mw. *ca.* 66,000) and has one tryptophan residue (in Subdomain IIA, Trp-214). Tryptophan is a relatively strong fluorescent amino acid [2] and is easily oxidized by oxidative stress, such as reactive oxygen species [3-7]. Because the redox potential of a one-electron oxidation of tryptophan is relatively low [8], electron transfer to a photoexcited molecule, of which the redox potential of a one-electron reduction is relatively low, also contributes to tryptophan oxidation. The purpose of this chapter is introduction of the application of HSA for phototherapy, specifically photodynamic therapy (PDT). A brief explanation of

PDT is presented in section 2. PDT is a cancer therapy using the photooxidation of biomacromolecules, including proteins [9-12]. To evaluate the PDT agents, protein-photodamaging activity and the binding interaction between agents and proteins are important. HSA is a usefull protein as a target biomacromolecule for these purposes. Because the X-ray structure and amino acid sequence of HSA have been clarified [1], speculation about its interaction with agents is relatively easy. Furthermore, oxidation damage of the tryptophan residue of HSA is easily evaluated via fluorometry. In this chapter, the application of HSA to develop PDT agents based on spectroscopic analysis of the binding interaction with agent molecules and oxidative damage of the tryptophan residue is introduced.

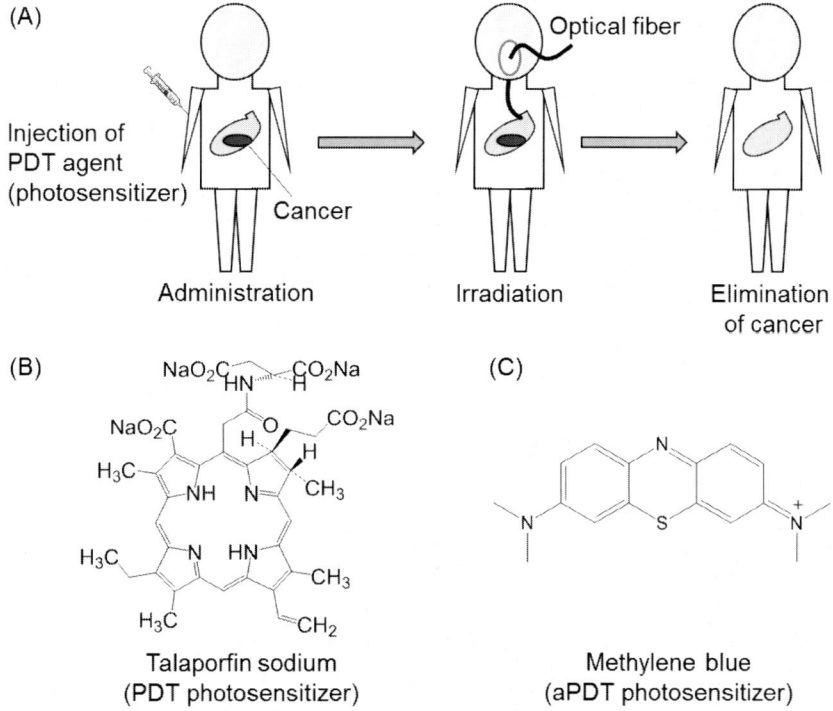

Figure 1. Scheme of the PDT process (A), an example of a clinically used PDT photosensitizer (B), and an example of an aPDT photosensitizer.

2. Protein Oxidation and Photodynamic Therapy

2.1. Photodynamic Therapy

The purpose of this review is the introduction of an evaluation method of PDT agents using HSA. Therefore, in this section, PDT is briefly reviewed. PDT, a promising treatment for cancer, is a medicinal application of photochemistry (Figure 1A) [9-12]. In general, porphyrin molecules are used for PDT agents (photosensitizers) (Figure 1B) [13, 14]. As PDT photosensitizers, they require no dark toxicity, however, photocytotoxicity under visible light irradiation. In general, PDT processes and their resulting events are as follows: administration of the photosensitizer by an injection, photoirradiation with an optical fiber after several hours or several days, and oxidative damage to the biomacromolecules in cancer cells through a photochemical reaction. The photosensitized reaction to the biomacromolecules is explained by the following two mechanisms: reactive oxygen species and electron transfer-mediated oxidation [15, 16]. Protein is the important targeting biomacromolecule, and HSA is used as a target protein model to evaluate the activity of PDT photosensitizers. Furthermore, antimicrobial PDT (aPDT) is an important medical and environmental application of photochemistry [17-21]. Phenothiazine dyes, such as methylene blue, are used for aPDT (Figure 1C). HSA can be also used to examine the photosensitizer activity of aPDT [22].

2.2. Singlet Oxygen Mechanism

In general, the PDT process can be explained by the generation of singlet oxygen ($^1O_2$), a reactive oxygen species, and the resulting oxidation of biomacromolecules [9-12]. The PDT mechanism through $^1O_2$ generation is called the Type II mechanism [15]. Although other reactive oxygen species such as superoxide, hydrogen peroxide, and hydroxyl radicals, are hardly produced by visible light irradiation, in which the photon energy is relatively small, $^1O_2$ can be easily produced by visible light energy. Singlet oxygen, usually generated through a Dexter energy transfer [23] from the photoexcited state of a photosensitizer, is the excited state of an oxygen molecule (Figure 2), and the following two states are possible; $^1\Delta_g$ and $^1\Sigma_g^+$; these have excitation energy of 0.98 eV and 1.63 eV above the ground state, respectively [24-26]. Since the lifetime of the $^1\Sigma_g^+$ type is very short (a few picoseconds), $^1\Delta_g$, the lower excited state of oxygen molecule is more important for the PDT mechanism. Amino acid residues of protein are oxidized by $^1O_2$, specifically, tryptophan [5, 6, 27, 28], tyrosine [29], histidine [30], cysteine [31], and methionine [32, 33] are important targets of $^1O_2$. In this chapter, the tryptophan residue of HSA is an important target to evaluate the PDT activity of photosensitizers.

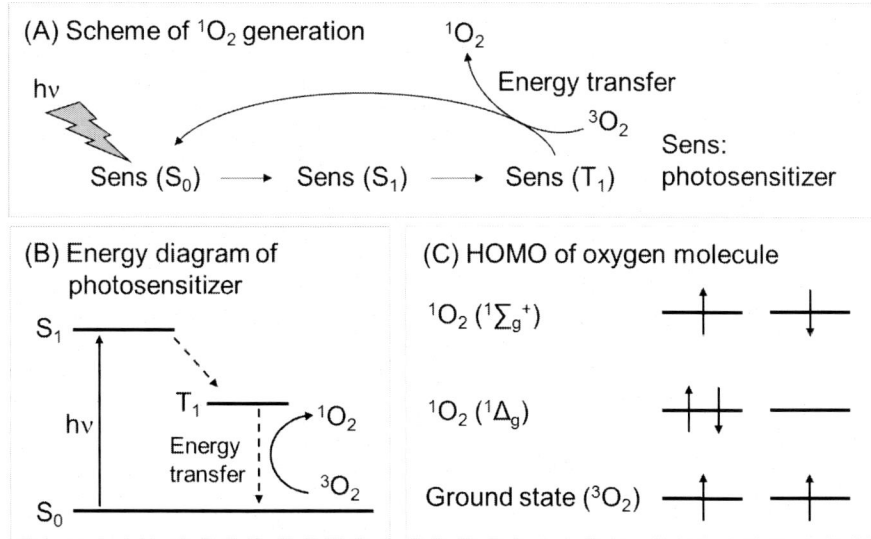

Figure 2. Scheme of photosensitized $^1O_2$ generation (A), energy diagram of $^1O_2$ generation (B), and the highest occupied molecular orbitals (HOMOs) of oxygen molecules (C).

2.3. Electron-Transfer Mechanism

Electron-transfer-mediated oxidation (the Type I mechanism) of protein is also an important mechanism of photocytotoxicity (Figure 3) [16]. Because the oxidation of biomolecules through electron transfer requires relatively high photon energy, the photooxidation of various biomolecules by an ultraviolet photosensitizer is explained by this mechanism [16]. However, visible light-induced biomolecule damage, including the PDT process, is also partly explained by electron transfer-mediated oxidation. An important problem of PDT is the hypoxic condition (low oxygen concentration) in cancer cells [34-36]. Therefore, a $^1O_2$-mediated PDT reaction is restricted in hypoxic cancer cells. Since the electron transfer-mediated mechanism does not directly depend on the oxygen concentration, electron transfer photosensitizers may solve the hypoxic problem to improve the PDT effect. The tryptophan residue of HSA, which has a lower redox potential of one-electron oxidation [8], is easily oxidized through electron transfer to photoexcited photosensitizers.

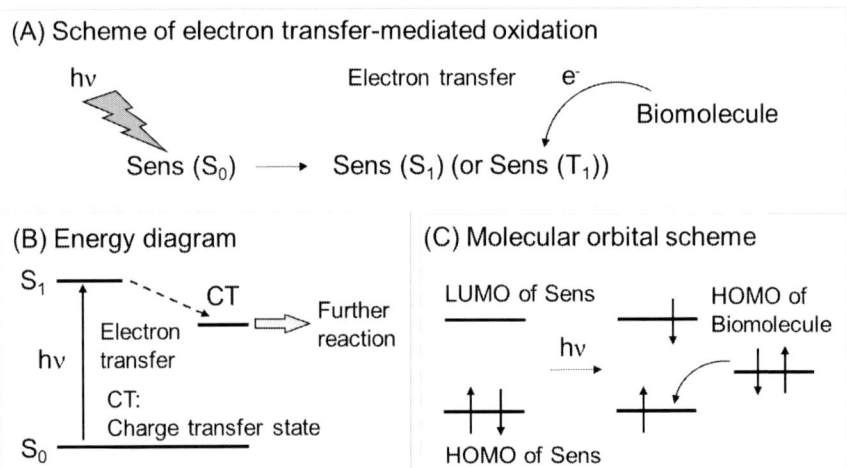

Figure 3. Electron transfer-mediated biomolecule oxidation scheme (A). energy diagram (B), and molecular orbital scheme (C).

2.4. Development of Photosensitizers

To develop the PDT photosensitizers, an evaluation of biomacro-molecule-damaging activity, specifically protein-damaging activity, is important. The binding interaction between the photosensitizers and protein is also an important characteristic. HSA, a water-soluble protein, is commercially available and useful to evaluate the activity of PDT photosensitizers. In this chapter, the binding interaction between HSA and the photosensitizer is described in section 3. The evaluation of photosensitizer activity using HSA is explained in section 4. Examples of the protocol for these evaluations are introduced in section 5.

3. ANALYSIS OF BINDING INTERACTION BETWEEN HSA AND PHOTOSENSITIZERS

3.1. Association Constant

The main regions of HSA agent binding sites are located in hydrophobic cavities called Subdomain IIA and Subdomain IIIA. PDT photosensitizers must absorb visible light, whereas the absorption band of HSA is only in the ultraviolet region. In general, the binding interaction with protein affects the electronic state of photosensitizers, resulting in the absorption spectral change. Therefore, an absorption technique is a useful tool to examine the binding interaction between HSA and photosensitizers (Figure 4). The apparent association constant (K_{ap}) under the assumption of 1:1 complex formation can be expressed by the following equation:

$$K_{ap} = \frac{[\text{HSA-Sens}]}{[\text{HSA}][\text{Sens}]} \tag{1}$$

where [HSA-Sens] is the concentration of the photosensitizer binding with HSA, [HSA] is the apparent concentration of the effective binding site of

HSA, and [Sens] is the concentration of non-binding photosensitizers. K_{ap} can be roughly calculated using the Benesi-Hildebrand equation [37, 38]:

$$\frac{[HSA]_0}{Abs-Abs_0} = \frac{[HSA]_0}{Abs_b-Abs_0} + \frac{1}{K_{ap}(Abs_b-Abs_0)} \quad (2)$$

where $[HSA]_0$ is the initial concentration of HSA and Abs, Abs_0, and Abs_b correspond to the observed absorbance of photosensitizers with HSA, the absorbance without HSA, and the absorbance of HSA-binding photosensitizers, respectively. In the plot of $[HSA]_0/(Abs - Abs_0)$ versus $[HSA]_0$, K_{ap} is given by the ratio of the slope to the intercept (Figure 5A).

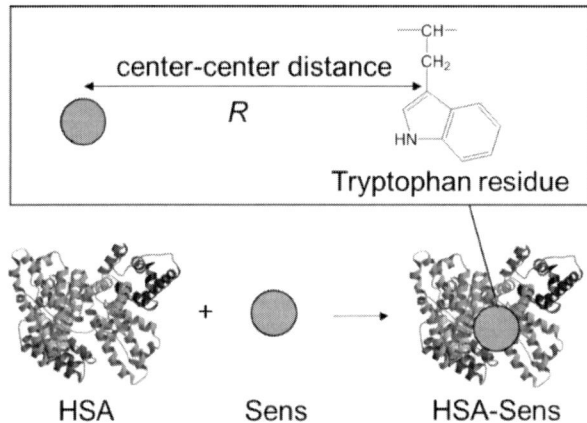

Figure 4. Scheme of binding interaction between HSA and the photosensitizer.

To obtain more exact values of K_{ap}, the following procedures can be performed [39, 40]. K_{ap} can be expressed by the following equation using the binding ratio (x):

$$K_{ap} = \frac{[Sens]_0 x}{([HSA]_0-[Sens]_0 x)[Sens]_0(1-x)} \quad (3)$$

where $[Sens]_0$ is the initial concentration of photosensitizers and the x is the binding ratio of photosensitizers with HSA. The x is calculated by the following equation using a certain K_{ap} value:

$$x = \frac{(K_{ap}[Sens]_0^2 + K_{ap}[Sens]_0[HSA]_0 + [Sens]_0)}{2K_{ap}[Sens]_0^2}$$
$$- \frac{\sqrt{(K_{ap}[Sens]_0^2 + K_{ap}[Sens]_0[HSA]_0 + [Sens]_0)^2 - 4K_{ap}^2[Sens]_0^3[HSA]_0}}{2K_{ap}[Sens]_0^2} \quad (4)$$

The observed absorbance of the photosensitizers at the corresponding wavelength, which is the same as the above-mentioned *Abs*, is expressed as follows:

$$Abs = Abs_b x + Abs_0 (1-x) \quad (5)$$

The above-mentioned values of Abs_b and x can be estimated from the analysis of the relationship between *Abs* and $[HSA]_0$ by the non-linear least squares method (Figure 5B). To obtain K_{ap}, the plots of *Abs* vs. $[HSA]_0$ are analyzed using equations (4) and (5).

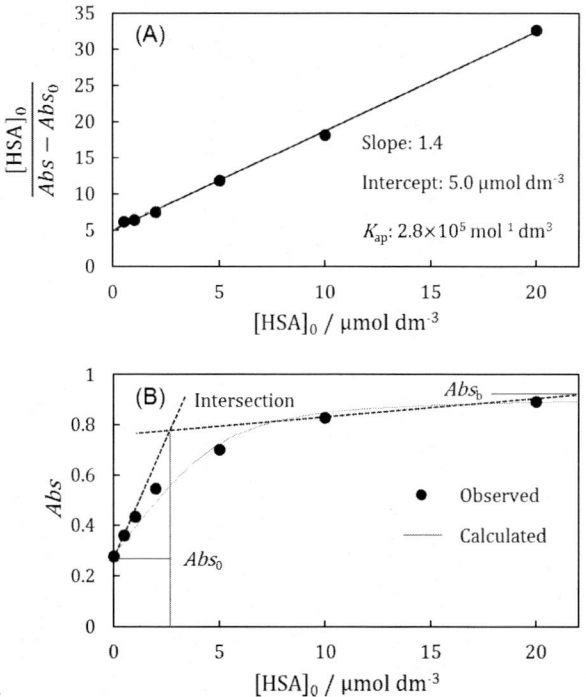

Figure 5. Example of the Benesi-Hildebrand plot (A) and the absorbance analysis (B).

Because HSA has several hydrophobic pockets [1], the binding ratio (the number of binding photosensitizers/HSA molecule) is also an important parameter. The intersection of two asymptotes of lower and higher HSA concentrations suggests the binding ratio [41]. A Job plot also supports the estimated binding ratio [41]. In the case of P(V)porphyrin photosensitizers, it was reported that the binding ratio of porphyrin: HSA becomes almost 2:1 [41] or 1:1 [42]. To analyze the 2:1 complex formation, the twice $[HSA]_0$ is used in the above-mentioned equations (1) - (4), because [HSA] is the effective concentration of the binding sites of HSA. Although the absorption spectral method is relatively easy, the circular dichroism (CD) spectral technique is strong tool for analyzing the binding interaction between the photosensitizers and proteins, specifically HSA [43-45]. In cases using the CD spectral method, a procedure similar to that for the absorption spectral method using the same equations is possible

3.2. Binding Mechanism and Thermodynamic Parameters

Hydrophobic interaction, electrostatic interaction, and the van der Waals force are considered to play the main roles in the binding interaction. To consider the binding mechanism, thermodynamic parameters, enthalpy changes (ΔH), entropy changes (ΔS), and Gibbs free energy (ΔG) are important. These parameters can be calculated from the temperature dependence of K_{ap} using the following equations derived from the van't Hoff equation and the definition of ΔG:

$$\Delta G = -RT \ln K_{ap} \qquad (6)$$

Table 1. Speculation of the driving force of binding interaction from the thermodynamic parameters

Driving force	ΔH	ΔS	ΔG
Hydrophobic interaction	Positive (+), small	Positive (+), large	Negative (-)
Electrostatic interaction	Negative (-), large	Positive (+)	Negative (-)
van der Waals force	Negative (-), small	Positive (+)	Negative (-)

and

$$\ln K_{\mathrm{ap}} = -\frac{\Delta H}{RT} + \frac{\Delta S}{R} \tag{7}$$

3.3. Binding Position

The binding position of photosensitizer molecules on HSA is also important information when considering the PDT activity. In general, X-ray diffraction (XRD) analysis is a strong tool. However, crystallization of the binding complex of HSA and the photosensitizer is difficult. Fluorescence resonance energy transfer (FRET), the mechanism of which has been explained by Förster [46], is a relatively simple method to determine the binding distance between the binding photosensitizer molecule and a certain fluorescent amino acid residue (Figure 4), though the position determination by this method includes a relatively large error. In general, tryptophan residue is used for this method. HSA has one tryptophan residue, and tryptophan emits in the ultraviolet region. Because the PDT photosensitizers must absorb visible light, photosensitizers can be excited by ultraviolet radiation, and the absorption band and tryptophan fluorescence spectrum overlap. Therefore, the fluorescence quenching of tryptophan by a visible light photosensitizer can be explained by the Förster energy transfer [46-49]. Although electron transfer-mediated quenching and/or the Dexter energy transfer [23] may be possible, these quenching processes are sensitive to distance between the donor and acceptor. Because long-distance quenching through Förster energy transfer is possible, the fluorescence quenching of tryptophan by photosensitizer can be explained by this energy transfer mechanism. The energy transfer rate depends on the distance between the donor and acceptor and the critical distance (R_0), at which the fluorescence quantum yield (Φ) of the donor becomes the half of that without an electron acceptor (Φ_0), as follows:

$$\frac{\Phi}{\Phi_0} = \frac{1}{1+\left(\frac{R}{R_0}\right)^{-6}} \tag{8}$$

where R is the distance between the donor and acceptor (Figure 4). The value of (Φ/Φ_0) is equal to the ratio of the fluorescence intensity of the donor with and without an acceptor. The R_0 is expressed by the following equation using the spectral overlap integral:

$$R_0^6 = \frac{9000c^4 \ln 10 \chi^2 \Phi_0}{128\pi^5 n^4 N_A} \int f(v)\varepsilon(v)\frac{dv}{v^4} \tag{9}$$

where c is the light velocity, χ is the orientation factor between donor and acceptor ($\chi^2 = 2/3$ at the random orientation), n is the refractive index of surroundings, N_A is the Avogadro constant, $f(v)$ is a normalized fluorescence spectrum of the donor, $\varepsilon(v)$ is the molar absorption coefficient of the acceptor, and v is the frequency of light. The binding position of the photosensitizer on HSA can be speculated using the fluorescence intensity of tryptophan with and without a photosensitizer, the fluorescence spectrum of tryptophan, and the absorption spectrum of the photosensitizer.

A docking simulation using commercially available software is also a convenient method to speculate the binding position [50, 51]. The binding distance estimated via the FRET method and a supportive docking simulation are relatively simple and effective tools to examine the binding position of the photosensitizer on HSA.

4. EVALUATION OF PHOTOSENSITIZED PROTEIN-DAMAGING ACTIVITY

4.1. Fluorometry of Photosensitized Tryptophan Damage

HSA emits intrinsic fluorescence in the ultraviolet region due to fluorescent amino acid residues [2, 41, 42]. Tryptophan residue emits an especially strong fluorescence at around 350 nm under illumination at around 300 nm. Tryptophan residue can be oxidized by $^1O_2$ and/or the electron extraction by photoexcited molecules, resulting in the diminishment of its fluorescence (Figure 6) [41, 42]. Therefore, fluorometry is a simple and useful method to evaluate the oxidative damage of the tryptophan residue of HSA. Simply, the concentration of damaged HSA ([Damaged HSA]) can be determined by the fluorescence intensity measurement using the following equation:

$$[\text{Damaged HSA}] = \frac{F_0 - F}{F_0} \times [\text{HSA}]_0 \tag{10}$$

where F_0 is the initial fluorescence intensity of the tryptophan residue of HSA, F is that of the photosensitized tryptophan residue, and $[\text{HSA}]_0$ is the initial concentration of HSA (same as the above-mentioned value). In addition, the photosensitized damage of free amino acids, such as tryptophan and/or tyrosine, can be examined with high-pressure liquid chromatography (HPLC) and/or liquid chromatography-mass spectrometer (LCMS) [52, 53].

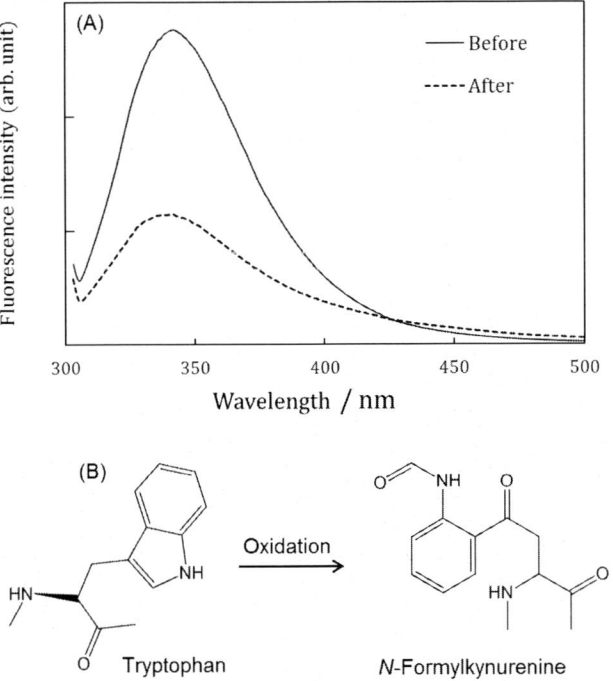

Figure 6. Fluorescence spectra of the tryptophan residue of HSA before and after photoirradiation with a photosensitizer (A). An example of the oxidized reaction of the tryptophan residue of HSA (B).

4.2. Examination of Protein-Damaging Mechanism

The mechanism of HSA damage photosensitized by PDT agents can be examined using this method. Under the assumption that the photosensitized protein damage is caused through $^1O_2$ generation and electron transfer-mediated oxidation, a scavenger effect can be used to evaluate their contributions [41, 42]. Sodium azide (NaN_3) is a strong physical scavenger of $^1O_2$ [54]. Observed protein damage in the presence of a sufficient concentration of NaN_3 is almost explained by the electron transfer-mediated mechanism and the ratio of inhibited protein damage with NaN_3 shows the contribution of the $^1O_2$-mediated mechanism.

The contributions of these mechanisms of protein damage, $^1O_2$ generation and electron transfer-mediated oxidation can be estimated using the above-mentioned method. However, complete scavenging of $^1O_2$ by NaN_3 is difficult, even though a large amount of NaN_3 is added to the sample solution. Furthermore, a high concentration of NaN_3 may quench the photoexcited state of the photosensitizer, resulting in inhibition of the electron transfer-mediated mechanism. To determine the exact values of the contributions of $^1O_2$ and electron transfer-mediated mechanisms, the following procedures can be used [41]. The total quantum yield of photosensitized protein damage (Φ_{TD}) can be expressed using their contributions as follows:

$$\Phi_{TD} = \Phi_\Delta \times \Phi_R + \Phi_{ET} \times \Phi_{DC} \tag{11}$$

where Φ_R is the quantum yield of the oxidation reaction of protein by the formed $^1O_2$, Φ_{ET} is the electron transfer quantum yield from the protein to the photoexcited photosensitizer, and Φ_{DC} is the quantum yield of the decomposition reaction from the charge transfer (CT) state. Consequently, $\Phi_\Delta \times \Phi_R$ and $\Phi_{ET} \times \Phi_{DC}$ in equation (11) indicate the absolute quantum yields of protein damage through $^1O_2$ generation and that of the electron transfer-mediated mechanism, respectively. These processes are shown in Figure 7. To analyze the experimental values simply, the following equations are useful [41]. The contributions of protein damage through $^1O_2$ generation (C_{AD}) and electron transfer (C_{ETD}) can be expressed as follows:

$$C_{AD} = \frac{\Phi_\Delta \times \Phi_R}{\Phi_{TD}} \times 100\% \tag{12}$$

$$C_{ETD} = \frac{\Phi_{ET} \times \Phi_{DC}}{\Phi_{TD}} \times 100\% \tag{13}$$

and

$$C_{AD} + C_{ETD} = 100\% \tag{14}$$

Due to the quenching effect of NaN_3 on the photoexcited state of the photosensitizer, determination of quenching rate coefficient is necessary to estimate these quantum yields. The quenching rate coefficient of the singlet excited (S_1) state by NaN_3 (k_{qf}) can be estimated via the fluorescence lifetime decrement using the Stern-Volmer plot (Figure 8A). For this analysis, the quenching rate coefficient of $^1O_2$ by NaN_3 ($k_{q\Delta}$) is also necessary. In certain cases, these quenching rates can be considered to be almost the same as the diffusion control reaction (k_{dif}), where k_{dif} becomes 7.4×10^9 mol^{-1} dm^3 s^{-1} in aqueous solution. To determine the exact value of $k_{q\Delta}$ in the presence of protein, the Stern-Volmer analysis of $^1O_2$ emission quenching by NaN_3 similar to that of k_{qf} is necessary (Figure 8B).

In the cases of P(V)porphyrin photosensitizers, their triplet excited (T_1) states are hardly quenched by NaN_3 [41]. Thus, the above-mentioned value of Φ_{TD} could be expressed as follows in the presence of NaN_3:

$$\Phi_{TD} = \Phi_{ET} \times \Phi_{DC} \times \frac{1}{1+\tau_f^0 k_{qf}[NaN_3]} + \Phi_\Delta \times \Phi_R \times \frac{1}{1+\tau_f^0 k_{qf}[NaN_3]} \times \frac{1}{1+\tau_\Delta^0 k_{q\Delta}[NaN_3]} \tag{15}$$

where τ_f^0 is the lifetime of the photoexcited photosensitizer without NaN_3 and τ_Δ^0 is that of $^1O_2$. To obtain the values of $\Phi_\Delta \times \Phi_R$ and $\Phi_{ET} \times \Phi_{DC}$, the relationship between Φ_{TD} (experimental data) and [NaN$_3$] is analyzed by a non-linear least-squares method using this equation (15) [41].

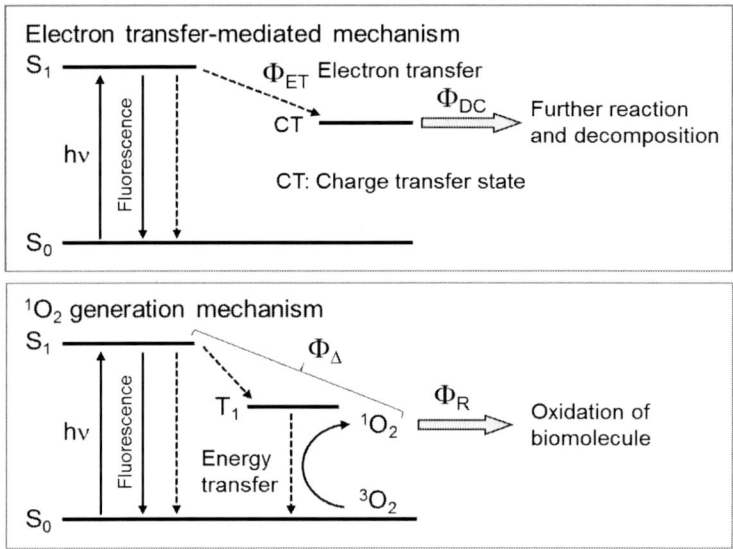

Figure 7. Relaxation process of the photoexcited photosensitizer and the protein-damaging mechanisms.

4.3. Kinetics of Protein Damage through Electron Transfer

Singlet oxygen oxidizes the amino acid residues of protein [5]. The rate constant of the reaction hardly depends on the kinds of photosensitizers. However, the kinetics of protein oxidation through electron transfer strongly depend on the kinds of photosensitizers [55-59]. In general, the redox potential of photosensitizers is an important parameter of the electron transfer-mediated protein oxidation. In general, the interaction with HSA stabilizes the photoexcited state of the photosensitizers. The S_1 state lifetime of the photosensitizers can be lengthened by inhibition of the vibrational deactivation of the S_1 state through an interaction with HSA. In the presence of a sufficient concentration of HSA, most photosensitizer molecules bind to HSA. Some of the photosensitizers are stabilized by HSA, and other molecules undergo the deactivation by a certain amino acid, which acts as an electron donor, such as tryptophan residue. Measurement of the fluorescence lifetime, which is equal to the S_1 state lifetime, is useful to evaluate the electron transfer rate [41, 42].

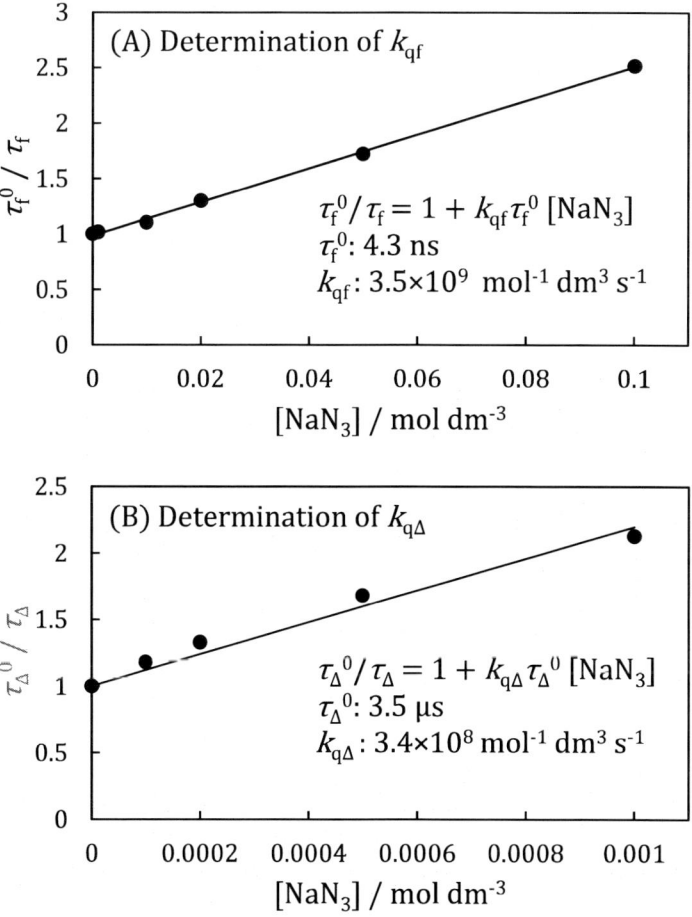

Figure 8. Examples of the Stern-Volmer plots to determine the kinetic parameters. A fluorescence quenching by NaN_3 (A). A quenching of $^1O_2$ emission by NaN_3 (B). The presented values are examples of P(V)porphyrins in an aqueous solution. τ_f and τ_Δ indicate the lifetimes of the photoexcited photosensitizer and $^1O_2$ in the presence of NaN_3, respectively.

If two fluorescence lifetime species are observed, the electron transfer rate constant (k_{ET}) from the tryptophan residue to the S_1 state of photosensitizers can be calculated by the following equation:

$$k_{ET} = \frac{1}{\tau_{st}} - \frac{1}{\tau_{lg}} \tag{16}$$

where τ_{st} is the fluorescence lifetime of shorter lifetime species and τ_{lg} is that of longer lifetime species.

In general, the T_1 state has relatively weak oxidation activity. In the cases of porphyrin P(V) complexes, the T_1 state lifetime increased in the presence of HSA [41]. This phenomenon can be explained by the inhibition of interaction with oxygen molecules. However, an interaction with HSA greatly increased the T_1 state lifetime of P(V)porphyrins under the anaerobic condition, as in the aerobic condition. These findings suggest that the T_1 state of P(V)porphyrins cannot oxidize HSA through electron transfer. Consequently, $^1O_2$ generation is the main mechanism of HSA oxidation from the T_1 state of P(V)porphyrins.

5. Protocol of Activity Evaluation of PDT Photosensitizers Using HSA

In this section, the practical procedures for examination of the binding interaction between HSA and photosensitizers and the evaluation of photodynamic activity of photosensitizers are introduced.

5.1. Interaction

As mentioned in section 3, the absorption spectral change of photosensitizers through an interaction with HSA is a convenient tool. For example, 10 µmol dm^{-3} HSA and several-micromolar photosensitizers are dissolved in a buffer solution, such as a 10 mmol dm^{-3} sodium phosphate buffer (around pH 7). The absorbance change of the photosensitizer in the visible light region is due to the interaction with HSA. The analysis of absorbance depending on the concentration of HSA using equations (1) - (5) gives us the binding constant between HSA and the photosensitizer. The thermodynamic parameters can be obtained via analysis of the temperature dependence of the binding constants. The driving force of the binding interaction can be speculated using the thermodynamic parameters (Table 1).

Using the FRET method, the binding position of photosensitizers on HSA can be speculated. For example, a sample solution containing 10 μmol dm^{-3} HSA and appropriate concentration of photosensitizer is photoexcited by ultraviolet light (for example, 298 nm). Because photoabsorption by a photosensitizer at the excitation wavelength decreases the photoabsorption by the tryptophan residue of HSA, the observed fluorescence intensity (I_{ob}) should be corrected using the absorbance ratio of HSA and the photosensitizer. Furthermore, the I_{ob} contains the fluorescence from non-binding HSA in addition to that from HSA binding with the photosensitizer molecule. Therefore, the I_{ob} can be expressed by the following relationship under the assumption that the absorption coefficient of tryptophan is independent of the photosensitizer:

$$I_{ob} \times \frac{(1-10^{-Abs_{Trp}})Abs_{Tot}}{(1-10^{-Abs_{Tot}})Abs_{Trp}} = I \times y + I_0 \times (1-y) \qquad (17)$$

where Abs_{Tot} is the absorbance of the sample solution at the excitation wavelength (298 nm), Abs_{Trp} is that of the HSA solution without a photosensitizer, I is the fluorescence intensity of 10 μmol dm^{-3} HSA binding with the photosensitizer, I_0 is that of 10 μmol dm^{-3} HSA without a photosensitizer, and y is the ratio of photosensitizer-binding HSA. The parameter y can be expressed by the following equation using the above-mentioned x (section 3.1, equation 5):

$$y = \frac{[Sens]_0 x}{[HSA]_0} \qquad (18)$$

For analysis using the FRET method, the fluorescence quantum yield ratio (equation 8) can be obtained by the following equation:

$$\frac{\Phi}{\Phi_0} = \frac{I}{I_0} = \frac{I_{ob}(1-10^{-Abs_{Trp}})Abs_{Tot}}{I_0(1-10^{-Abs_{Tot}})Abs_{Trp} \times y} - \frac{1}{y} + 1 \quad (19)$$

After measurements of the absorption spectrum of the appropriate concentration of photosensitizer and the fluorescence spectrum of the tryptophan residue of HSA at 298 nm excitation, the spectral overlap can be calculated using equation (9). Using these parameters with equation 8, the distance between the tryptophan residue of HSA and the photosensitizer can be speculated. If the fluorescence lifetime measurement of HSA is possible, this procedure is simpler. Using the fluorescence lifetime ratio of the tryptophan residue, equation (8) can be changed as follows:

$$\frac{\tau(Trp)}{\tau(Trp)_0} = \frac{1}{1+\left(\frac{R}{R_0}\right)^{-6}} \quad (20)$$

where $\tau(Trp)$ is the fluorescence lifetime of the tryptophan residue quenched by the photosensitizer and $\tau(Trp)_0$ is that without photosensitizer.

5.2. Protein-Damaging Activity

Oxidative damage of the tryptophan residue of HSA is evaluated via fluorometry of intrinsic fluorescence of tryptophan. For example, 10 μmol dm^{-3} HSA and several-micromolar photosensitizers are dissolved in a buffer solution such as a 10 mmol dm^{-3} sodium phosphate buffer (around pH 7). Before and after photoirradiation of the sample solution, the fluorescence intensity (or spectrum) is measured by the photoexcitation at 298 nm (Figure 9). Protein damage can be evaluated using the fluorescence intensity by the equation (10). The value of Φ_{TD}, which is the quantum yield of HSA damage described in section 4.2, can be also determined using this method. In general, a linear relationship between [Damaged HSA] and the irradiation time is observed during an initial reaction period. Consequently, the slope of the time profile indicates the rate (R_{ox}, unit; mol s^{-1}) of the photosensitized oxidation reaction. To calculate the Φ_{TD} value, the spectrum (or intensity) of the excitation light source should be determined. If the wavelength width of the light source is narrow, the light intensity (*LI*, unit; W cm^{-2}) can be considered constant. In this case, the Φ_{TD} can be expressed as follows using the photon fluence absorbed by the photosensitizer (*PF*, unit; mol s^{-1}):

$$\Phi_{TD} = \frac{R_{OX}}{PF} \tag{21}$$

PF can be estimated using the following equation [60]:

$$PF = \frac{LI \times S \times \lambda}{h \times c \times N_A} \times \left(1 - 10^{\varepsilon(LS) \times M \times l}\right) \tag{22}$$

where *S* is the irradiation area of the sample (unit; cm^2), λ is the wavelength of irradiation light (unit; m), *h* is the Plank constant (6.626 × 10^{-34} J s), *c* is the above-mentioned (in equation 9) light velocity (2.998 × 10^8 m s^{-1}), N_A is the above-mentioned (in equation 9) Avogadro constant (6.02 × 10^{23} mol^{-1}), ε(LS) is the molar absorption coefficient at irradiation wavelength (unit; mol^{-1} dm^3 cm^{-1}), *M* is the concentration of the photosensitizer molecule (unit; mol dm^{-3}), and *l* is the light-pass length of the sample solution (unit; cm). When the wavelength width of the light source is not negligible, the *PF* value is calculated via the following equation using the spectrum of light source (*LS*(λ), unit; W cm^{-2} nm^{-1}) [60] (Figure 10):

$$PF = \frac{S \times \lambda}{h \times c \times N_A} \int \{LS(\lambda) \times (1 - 10^{\varepsilon(LS) \times M \times l})\} d\lambda \tag{23}$$

where $d\lambda$ is considered to be have the unit (nm). Using equations (11) - (15), the mechanism of photosensitized protein damage can be analyzed. In general, the values of $\Phi_\Delta \times \Phi_R$ and $\Phi_{ET} \times \Phi_{DC}$ in equation (15) can be obtained by a non-linear least-squares method using experimental data [41].

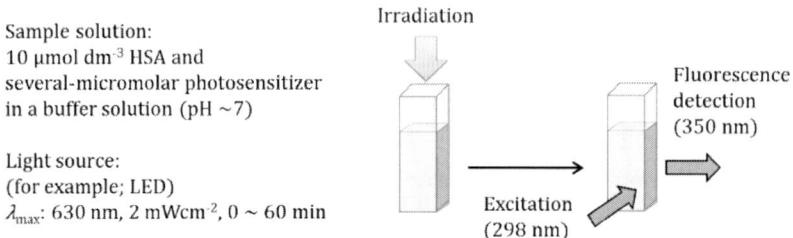

Figure 9. Scheme of the evaluation of photosensitized HSA damage via fluorometry.

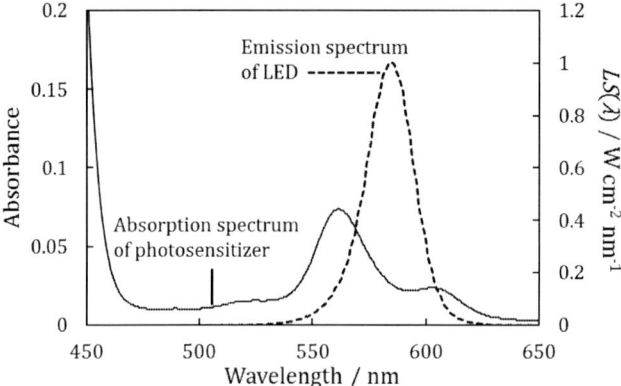

Figure 10. Example of the overlapping of emission spectrum of the light source and absorption spectrum of a photosensitizer.

SUMMARY

The absorption spectral change of a photosensitizer in the visible light region can be used to analyze the binding interaction with HSA. The absorbance change is used to determine the association constant between HSA and the photosensitizer. Its temperature dependence gives us the thermodynamic parameters. The FRET method is also used to speculate the binding distance between the tryptophan residue of HSA and the photosensitizer. Fluorometry of the tryptophan can be used to evaluate the HSA damage and PDT activity of photosensitizers. Using a scavenger for reactive oxygen species, the protein-damaging mechanism can be examined.

In the last section, these protocols were introduced. Because HSA is comercially available and can be used easily, various applications, including those in the medicinal field, are expected.

REFERENCES

[1] He, X. M. & Carter, D. C. (1992). Atomic structure and chemistry of human serum albumin. *Nature*, *358*, 209–215.

[2] Zelent, B., Bialas, C., Gryczynski, I., Chen, P., Chib, R., Lewerissa, K., Corradini, M. G., Ludescher, R. D., Vanderkooi, J. M. & Matschinsky, F. M. (2017). Tryptophan fluorescence yields and lifetimes as a probe of conformational changes in human glucokinase. *J. Fluoresc.*, *27*, 1621–1631.

[3] Davies, M. J. (2016). Protein oxidation and peroxidation. *Biochem. J.*, *473*, 805–825.

[4] Gracanin, M., Hawkins, C. L., Pattison, D. I. & Davies, M. J. (2009). Singlet-oxygen-mediated amino acid and protein oxidation: formation of tryptophan peroxides and decomposition products. *Free Radic. Biol. Med.*, *47*, 92–102.

[5] Jensen, R. L., Arnbjerg, J. & Ogilby, P. R. (2012). Reaction of singlet oxygen with tryptophan in proteins: a pronounced effect of the local environment on the reaction rate. *J. Am. Chem. Soc.*, *134*, 9820–9826.

[6] Ehrenshaft, M., Deterding, L. J. & Mason, R. P. (2015). Tripping up Trp: Modification of protein tryptophan residues by reactive oxygen species, modes of detection, and biological consequences. *Free Radic. Biol. Med.*, *89*, 220–208.

[7] Carroll, L., Pattison, D. I., Davies, J. B., Anderson, R. F., Lopez-Alarcon, C. & Davies, M. J. (2018). Superoxide radicals react with peptide-derived tryptophan radicals with very high rate constants to give hydroperoxides as major products. *Free Radic. Biol. Med.*, *118*, 126–136.

[8] Nan, C. G., Feng, Z. Z., Li, W. X., Ping, D. J. & Qin, C. H. (2002). Electrochemical behavior of tryptophan and its derivatives at a glassy

carbon electrode modified with hemin. *Analytica Chim. Acta*, *452*, 245–254.

[9] Dolmans, D. E. J. G. J., Fukumura, D. & Jain, R. K. (2003). Photodynamic therapy for cancer. *Nat. Rev. Cancer*, *3*, 380–387.

[10] Collins, H. A., Khurana, M., Moriyama, E. H., Mariampillai, A., Dahlstedt, E., Balaz, M., Kuimova, M. K., Drobizhev, M., Yang, V. X. D., Phillips, D., Rebane, A., Wilson, B. C. & Anderson, H. L. (2008). Blood-vessel closure using photosensitizers engineered for two-photon excitation. *Nat. Photonics,* *2*, 420–424.

[11] Soriano, J., Mora-Espí, I., Alea-Reyes, M. E., Pérez-García, L., Barrios, L., Ibáñez, E. & Nogués, C. (2017). Cell death mechanisms in tumoral and non-tumoral human cell lines triggered by photodynamic treatments: apoptosis, necrosis and parthanatos. *Sci. Rep.*, *7*, Article number: 41340.

[12] Chilakamarthi, U. & Giribabu, L. (2017). Photodynamic therapy: past, present and future. *The Chem. Rec.*, *17*, 1–29.

[13] Foster, T. H., Gibson, S. L. & Raubertas, R. F. (1996). Response of Photofrin®-sensitised mesothelioma xenografts to photodynamic therapy with 514 nm light. *British J. Cancer*, *73*, 933–936.

[14] Wang, S., Bromley, E., Xu, L., Chen, J. C. & Keltner, L. (2010). Talaporfin sodium. *Expert Opin. Pharmacother.* *11*, 133–140.

[15] Foote, C. S. (1991). Definition of Type I and Type II photosensitized oxidation. *Photochem. Photobiol.*, *54*, 659.

[16] Hirakawa, K. (2008). DNA damage through photo-induced electron transfer and photosensitized generation of reactive oxygen species, ed. by Kimura H. and Suzuki A. in *"New Research on DNA Damage"*, Chapter 9, pp. 197-219, Nova Science Publishers Inc., New York.

[17] Wainwright, M. & McLean, A. (2017). Rational design of phenothiazinium derivatives and photoantimicrobial drug discovery. *Dyes Pigment*, *136*, 590–600.

[18] Diogo, P., Gonçalves, T., Palma, P. & Santos, J. M. (2015). Photodynamic antimicrobial chemotherapy for root canal system asepsis: a narrative literature review. *Int. J. Dent.*, *2015*, 269205.

[19] Souza, E., Medeiros, A. C., Gurgel, B. C. & Sarmento, C. (2016). Antimicrobial photodynamic therapy in the treatment of aggressive periodontitis: a systematic review and meta-analysis. *Lasers Med. Sci.*, *31*, 187–196.

[20] Oruba, Z., Łabuz, P., Macyk, W. & Chomyszyn-Gajewska, M. (2015). Antimicrobial photodynamic therapy-A discovery originating from the pre-antibiotic era in a novel periodontal therapy. *Photodiagnosis Photodyn. Ther.*, *12*, 612–618.

[21] Tim, M. (2015). Strategies to optimize photosensitizers for photodynamic inactivation of bacteria. *J. Photochem. Photobiol. B*, *150*, 2–10.

[22] Hirakawa, K. & Ishikawa, T. (2017). Phenothiazine dyes photosensitize protein damage through electron transfer and singlet oxygen generation. *Dyes Pigment.*, *142*, 183–188.

[23] Dexter, D. L. (1953). A theory of sensitized luminescence in solids. *J. Chem. Phys.*, *21*, 836–850.

[24] DeRosa, M. C. & Crutchley, R. J. (2002). Photosensitized singlet oxygen and its applications. *Coord. Chem. Rev.*, *233-234*, 351–371.

[25] Schweitzer, C. & Schmidt, R. (2003). Physical mechanisms of generation and deactivation of singlet oxygen. *Chem. Rev.*, *103*, 1685–1758.

[26] Krasnovsky, Jr. A. A. (2018). Singlet molecular oxygen: Early history of spectroscopic and photochemical studies with contributions of A. N. Terenin and Terenin's school. *J. Photochem. Photobiol. A*, *354*, 11–24.

[27] Gracanin, M., Hawkins, C. L., Pattison, D. I. & Davies, M. J. (2009). Singlet-oxygen-mediated amino acid and protein oxidation: Formation of tryptophan peroxides and decomposition products. *Free Radic. Biol. Med.*, *47*, 92–102.

[28] Ronsein, G. E., Oliveira, M. C. B., Miyamoto, S., Medeiros, M. H. G. & Di Mascio, P. (2018). Tryptophan oxidation by singlet molecular oxygen [O_2 ($^1\Delta_g$)]: mechanistic studies using ^{18}O-labeled hydroperoxides, mass spectroscopy and light emission measurements. *Chem. Res. Toxicol.*, *21*, 1271–1283.

[29] Wright, A., Bubb, W. A., Hawkins, C. L. & Davies, M. J. (2002). Singlet oxygen-mediated protein oxidation: evidence for the formation

of reactive side chain peroxides on tyrosine residues. *Photochem. Photobiol.*, *76*, 35–46.

[30] Agon, V. V., Bubb, W. A., Hawkins, C. L. & Davies, M. J. (2006). Sensitizer-mediated photooxidation of histidine residues: evidence for the formation of reactive side-chain peroxides. *Free Radic. Biol. Med.*, *40*, 698–710.

[31] Chu, C., Erickson, P. R., Lundeen, R. A., Stamatelatos, D., Alaimo, P. J., Latch, D. E. & McNeill, K. (2016). Photochemical and nonphotochemical transformations of cysteine with dissolved organic matter. *Environ. Sci. Technol.*, *50*, 6363–6373.

[32] Liu, F. & Liu, J. (2015). Oxidation dynamics of methionine with singlet oxygen: effects of methionine ionization and microsolvation. *J. Phys. Chem. B*, *119*, 8001–8012.

[33] Liu, F., Lu, W., Yin, X. & Liu, J. (2016). Mechanistic and kinetic study of singlet O_2 oxidation of methionine by on-line electrospray ionization mass spectrometry. *J. Am. Soc. Mass. Spectrom.*, *27*, 59–72.

[34] Bratasz, A., Kulkarni, A. C. & Kuppusamy, P. (2007). A highly sensitive biocompatible spin probe for imaging of oxygen concentration in tissues. *Biophys. J.*, *92*, 2918–2925.

[35] Brahimi-Horn, C., Berra, E. & Pouysségur, J. (2011). Hypoxia: the tumor's gateway to progression along the angiogenic pathway. *Trends Cell Biol.*, *11*, S32–S36.

[36] Prior, S., Kim, A., Yoshihara, T., Tobita, S., Takeuchi, T. & Higuchi, M. (2014). Mitochondrial respiratory function induces endogenous hypoxia. *PLos One*, *9*, e88911.

[37] Benesi, H. A. & Hildebrand, J. H. (1949). A spectrophotometric investigation of the interaction of iodine with aromatic hydrocarbons. *J. Am. Chem. Soc.*, *71*, 2703–2707.

[38] Pyle, A. M., Rehman, J. P., Meshoyrer, R., Kumar, C. V., Turro, N. J. & Barton, J. K. (1989). Mixed-ligand complexes of ruthenium(II): factors governing binding to DNA. *J. Am. Chem. Soc.*, *111*, 3051–3058.

[39] Hirakawa, K. & Nakajima, S. (2014). Effect of DNA microenvironment on photosensitized reaction of water soluble cationic porphyrins. *Recent Adv. DNA Gene Seq.*, *8*, 35–43.

[40] Hirakawa, K., Taguchi, M. & Okazaki, S. (2015). Relaxation process of photoexcited meso-naphthylporphyrins while interacting with DNA and singlet oxygen generation. *J Phys Chem B., 119*, 13071–13078.

[41] Hirakawa, K., Umemoto, H., Kikuchi, R., Yamaguchi, H., Nishimura, Y., Arai, T., Okazaki, S. & Segawa, H. (2015). Determination of singlet oxygen and electron transfer mediated mechanisms of photosensitized protein damage by phosphorus(V)porphyrins. *Chem. Res. Toxicol., 28*, 262–267.

[42] Hirakawa, K., Ouyang, D., Ibuki, Y., Hirohara, S., Okazaki, S., Kono, E., Kanayama, N., Nakazaki, J. & Segawa, H. (2018). Photosensitized protein-damaging activity, cytotoxicity, and antitumor effects of P(V)porphyrins using long-wavelength visible light through electron transfer. *Chem. Res. Toxicol., 31*, 371–379.

[43] Nishijima, M., Pace, T. C. S., Nakamura, A., Mori, T., Wada, T., Bohne, C. & Inoue, Y. (2007). Supramolecular photochirogenesis with biomolecules. mechanistic studies on the enantiodifferentiation for the photocyclodimerization of 2-anthracenecarboxylate mediated by bovine serum albumin. *J. Org. Chem., 72*, 2707–2715.

[44] Nishijima, M., Wada, T., Mori, T., Pace, T. C. S., Bohne, C. & Inoue, Y. (2007). Highly enantiomeric supramolecular [4 + 4] photocyclodimerization of 2-anthracenecarboxylate mediated by human serum albumin. *J. Am. Chem. Soc., 129*, 3478–3479.

[45] Pace, T. C. S., Nishijima, M., Wada, T., Inoue, Y. & Bohne, C. (2009). Photophysical studies on the supramolecular photochirogenesis for the photocyclodimerization of 2-anthracenecarboxylate within human serum albumin. *J. Phys. Chem. B, 113*, 10445–10453.

[46] Förster, T. (1948). Zwischenmolekulare energiewanderung und fluoreszenz. [Intermolecular energy migration and fluorescence]. *Ann. Physik, 437*, 55–75.

[47] Abou-Zied, O. K. & Al-Shihi, O. I. K. (2008). Characterization of subdomain IIA binding site of human serum albumin in its native, unfolded, and refolded states using small molecular probes. *J. Am. Chem. Soc., 130*, 10793–10801.

[48] Teunissen, A. J. P., Pérez-Medina, C., Meijerink, A. & Mulder, W. J. M. (2018). Investigating supramolecular systems using Förster resonance energy transfer. *Chem. Soc. Rev.*, *47*, 7027–7044.

[49] Rabbani, G., Lee, E. J., Ahmad, K., Baig, M. H. & Choi, I. (2018). Binding of tolperisone hydrochloride with human serum albumin: effects on the conformation, thermodynamics, and activity of HSA. *Mol. Pharm.*, *15*, 1445–1456.

[50] Kuang, Z. K., Feng, S. Y., Hu, B., Wang, D., He, S. B. & Kong, D. X. (2016). Predicting subtype selectivity of dopamine receptor ligands with three-dimensional biologically relevant spectrum. *Chem. Biol. Drug Design*, *88*, 859–872.

[51] Chen, Y., Zhou, Y., Chen, M., Xie, B., Yang, J., Chen, J. & Sun, Z. (2018). Isorenieratene interaction with human serum albumin: Multi-spectroscopic analyses and docking simulation. *Food Chem.*, *258*, 393–399.

[52] Hirakawa, K., Fukunaga, N., Nishimura, Y., Arai, T. & Okazaki, S. (2013). Photosensitized protein damage by dimethoxyphosphorus(V) tetraphenylporphyrin. *Bioorg. Med. Chem. Lett.*, *23*, 2704–2707.

[53] Ouyang, D. & Hirakawa, K. (2017). Photosensitized enzyme deactivation and protein oxidation by axial-substituted phosphorus(V) tetraphenylporphyrins. *J. Photochem. Photobiol. B*, *175*, 125–131.

[54] Li, M. Y., Cline, C. S., Koker, E. B., Carmichael, H. H., Chignell, C. F. & Bilski, P. (2001). Quenching of singlet molecular oxygen ($^1O_2$) by azide anion in solvent mixtures. *Photochem. Photobiol.*, *74*, 760–764.

[55] Marcus, R. A. (1956). On the theory of oxidation-reduction reactions involving electron transfer. I. *J. Chem. Phys.*, *24*, 966–978.

[56] Marcus, R. A. & Sutin, N. (1985). Electron transfers in chemistry and biology. *Biochim. Biophys. Acta*, *811*, 265–322.

[57] Wasielewski, M. R. (1992). Photoinduced electron transfer in supramolecular systems for artificial photosynthesis. *Chem. Rev.*, *92*, 435–461.

[58] Hirakawa, K. & Yoshioka, T. (2015). Photoexcited riboflavin induces oxidative damage to human serum albumin, *Chem. Phys. Lett., 634,* 221–224.
[59] Hirakawa, K. & Morimoto, S. (2016). Electron transfer mediated decomposition of folic acid by photoexcited dimethoxophosphorus (V)porphyrin. *J. Photochem. Photobiol. A, 318,* 1–6.
[60] Hirakawa, K. (2009). Fluorometry of singlet oxygen generated via a photosensitized reaction using folic acid and methotrexate. *Anal. Bioanal. Chem., 393,* 999–1005.

BIOGRAPHICAL SKETCH

Kazutaka Hirakawa

Affiliation: Shizuoka University

Education:
Graduate School of Arts and Sciences, The University of Tokyo (PhD)

Research and Professional Experience:

- 2000-2004: Mie University School of Medicine,
- 2004- Faculty of Engineering, Shizuoka University
- 2015- Applied Chemistry and Biochemical Engineering Course, Department of Engineering, Graduate School of Integrated Science and Technology, Shizuoka University

Professional Appointments: Professor (2016-)

Honors:

- CSJ Presentation Award 2007 (2007, form The Chemical Society of Japan)
- Award from Japanese Society for Photomedicine and Photobiology (2009)
- IJRC Award (2010, from Innovation and Joint Research Center, Shizuoka University)
- MMS Award (2016, from Tanaka Kikinzoku Memorial Foundation)

Publications from the Last 3 Years:

[1] Kazutaka, Hirakawa., Tetuya, Kaneko. & Naoki, Toshima. "Kinetics of spontaneous bimetallization between silver and noble metal nanoparticles", *Chemistry - An Asian Journal*, *13*, 1892–1896, 2018, (Front Cover).

[2] Myeong, Eun Heo., Young-Ae, Lee., Kazutaka, Hirakawa., Shigetoshi, Okazaki., Seog, K. Kim. & Dae, Won Cho. "Sequence selective photoinduced electron transfer of a pyrene–porphyrin dyad to DNA", *Physical Chemistry Chemical Physics*, *20*, 16386–16392, 2018.

[3] Kazutaka, Hirakawa., Dongyan, Ouyang., Yuko, Ibuki., Shiho, Hirohara., Shigetoshi, Okazaki., Eiji, Kono., Naohiro, Kanayama., Jotaro, Nakazaki. & Hiroshi, Segawa. "Photosensitized protein-damaging activity, cytotoxicity, and antitumor effects of P(V)porphyrins using long-wavelength visible-light through electron transfer", *Chemical Research in Toxicology*, *31*, 371–379, 2018.

[4] Kazutaka, Hirakawa. & Atsushi, Murata. "Photosensitized oxidation of nicotinamide adenine dinucleotide by diethoxyphosphorus(V)tetraphenylporphyrin and its fluorinated derivative: Possibility of chain reaction", *Spectrochimica Acta Part A: Molecular and Biomolecular Spectroscopy*, *188*, 640–646, 2018.

[5] Shiho, Ohnishi., Yusuke, Hiraku., Keishi, Hasegawa., Kazutaka, Hirakawa., Shinji, Oikawa., Mariko, Murata. & Shosuke, Kawanishi.

"Mechanism of oxidative DNA damage induced by metabolites of carcinogenic naphthalene", *Mutation Research - Genetic Toxicology and Environmental Mutagenesis*, *827*, 42–49, 2018.

[6] Kazutaka, Hirakawa. & Takaya, Ishikawa. "Phenothiazine dyes photosensitize protein damage through electron transfer and singlet oxygen generation", *Dyes and Pigments*, *142*, 183–188, 2017.

[7] Hyun, Suk Lee., Ji, Hoon Han., Jin, Ha Park., Myeong, Eun Heo., Kazutaka, Hirakawa., Seog, K. Kim. & Dae, Won Cho. "Relationship between the photoinduced electron transfer and binding modes of a pyrene–porphyrin to DNA", *Physical Chemistry Chemical Physics*, *19*, 27123–27131, 2017.

[8] Dongyan, Ouyang. & Kazutaka, Hirakawa. "Photosensitized enzyme deactivation and protein oxidation by axial-substituted phosphorus(V) tetraphenylporphyrins", *Journal of Photochemistry and Photobiology, B: Biology*, *175*, 125–131, 2017.

[9] Kazutaka, Hirakawa. & Hiroshi, Segawa. "Multi-step intramolecular excitation energy transfer in dendritic pyrene-phosphorus(V)porphyrin heptads", *Journal of Luminescence*, *179*, 457–462, 2016.

[10] Kazutaka, Hirakawa. & Shu, Morimoto. "Electron transfer mediated decomposition of folic acid by photoexcited dimethoxophosphorus(V) porphyrin," *Journal of Photochemistry and Photobiology, A: Chemistry*, *318*, 1–6, 2016.

[11] Dongyan, Ouyang., Shiori, Inoue., Shigetoshi, Okazaki. & Kazutaka, Hirakawa. "Tetrakis(*N*-methyl-*p*-pyridinio)porphyrin and its zinc complex can photosensitize damage of human serum albumin through electron transfer and singlet oxygen generation," *Journal of Porphyrins and Phthalocyanines*, *20*, 813–821, 2016.

[12] Kazutaka, Hirakawa., Makoto, Taguchi. & Shigetoshi, Okazaki. "Relaxation process of photoexcited *meso*-naphthylporphyrins while interacting with DNA and singlet oxygen generation," *The Journal of Physical Chemistry B*, *119*, 13071–13078, 2015.

[13] Kazutaka, Hirakawa., Hironobu, Umemoto., Ryo, Kikuchi., Hiroki, Yamaguchi., Yoshinobu, Nishimura., Tatsuo, Arai., Shigetoshi, Okazaki. & Hiroshi, Segawa. "Determination of singlet oxygen and

electron transfer mediated mechanisms of photosensitized protein damage by phosphorus(V)porphyrins," *Chemical Research in Toxicology*, *28*, 262–267, 2015.

[14] Kazutaka, Hirakawa. & Takuto, Yoshioka. "Photoexcited riboflavin induces oxidative damage to human serum albumin," *Chemical Physics Letters*, *634*, 221–224, 2015.

[15] Kazutaka, Hirakawa. & Hiroki, Ito. "Rhodamine-6G can photosensitize folic acid decomposition through electron transfer," *Chemical Physics Letters*, *627*, 26–29, 2015.

[16] Kazutaka, Hirakawa., Shunsuke, Aoki., Hiroyuki, Ueda., Dongyan, Ouyang. & Shigetoshi, Okazaki. "Photochemical property and photodynamic activity of tetrakis(2-naphthyl) porphyrin phosphorus (V) complex," *Rapid Communication in Photoscience*, *4*, 37–40, 2015.

[17] Takashi, Ohtsuki., Shunya, Miki., Shouhei, Kobayashi., Tokuko, Haraguchi., Eiji, Nakata., Kazutaka, Hirakawa., Kensuke, Sumita., Kazunori, Watanabe. & Shigetoshi, Okazaki. "The molecular mechanism of photochemical internalization of cell penetrating peptide-cargo-photosensitizer conjugates," *Scientific Reports*, DOI: 10.1038/srep18577, 1–9, 2015.

[18] Reiko, Makita., Mitsuji, Yamashita., Mayumi, Yamaoka., Michio, Fujie., Satoki, Nakamura., Tatsuo, Oshikawa., Junko, Yamashita., Manabu, Yamada., Kazuhide, Asai., Takuya, Suyama., Mitsuru, Kondo., Hiroko, Hasegawa., Yoshimitsu, Okita., Kazutaka, Hirakawa., Mitsuo, Toda., Kazunori, Ohnishi. & Haruhiko, Sugimura. "Novel multiple type molecular targeted antitumor agents: preparation and preclinical evaluation of low-molecular-weight phospha sugar derivatives," *Phosphorus, Sulfur, and Silicon and the Related Elements*, *190*, 733–740, 2015.

[19] Dongyan, Ouyang. & Kazutaka, Hirakawa. "Photosensitized oxidative damage of human serum albumin by water-soluble dichlorophosphorus(V) tetraphenylporphyrin," *Rapid Communication in Photoscience*, *4*, 41–44, 2015.

In: Human Serum Albumin
Editor: Dianne Cohen

ISBN: 978-1-53614-787-2
© 2019 Nova Science Publishers, Inc.

Chapter 4

MAGNETIC RESONANCE IMAGING AND SPECTROSCOPY FOR THE DETERMINATION OF HUMAN SERUM ALBUMIN

Dorota Bartusik-Aebisher[*], *David Aebisher and Adrian Truszkiewicz*

[1]Faculty of Medicine, University of Rzeszów, Rzeszów, Poland

ABSTRACT

The aim of this chapter is to review the use of Magnetic Resonance Imaging and Spectroscopy for the determination of Human Serum Albumin structure, drug binding and in vivo activity.

PubMed, Embase, the Cochrane Library, Elsevier, Wiley, and Ovid were searched for randomized controlled trials and prospective studies. Also, we discuss studies which describe drug modifications using Human Serum Albumin.

---

[*] Corresponding Author Email: dbartusik-aebisher@ur.edu.pl.

Keywords: Human Serum Albumin, monitoring, Magnetic Resonance Imaging, Magnetic Resonance Spectroscopy

MAGNETIC RESONANCE IMAGING AND SPECTROSCOPY

An important tool in any therapy is quantitative monitoring of the response to treatment. There is a potential in using magnetic resonance imaging and magnetic resonance spectroscopy for monitoring of therapy as an early response to treatment on the morphological level. Drugs binding to human serum albumin tubulin are known to disrupt the dynamics of cancer cell growth. Advances in synthetic methods, coupled with a continued effort to identify cell surface receptors are listed in Table 1 below.

In a study by Lenora et al. 2017, relaxivity was measured for a series of EuII adducts with β-cyclodextrins, poly-β-cyclodextrins, and human serum albumin (Lenora et al. 2017). MRI measurememnts with iron oxide nanoparticles conjugated with human serum albumin (Erdal et al. 2018), and 99mTc-galactosyl albumin were performed (Morine et al. 2017; Aurich et al. 2017). In another study, nanoparticles with human serum albumin and sialic acid were imaged by using MRI techniques (Nasr et al. 2018; Shamsutdinova et al. 2018). Nanoparticles consisting of human serum labumin and MnO_2-Ce6 were imaged with MRI and then used to treat orthotopic bladder cancer by PDT (Lin et al. 2018). In another study, a single emulsion method was utilized to fabricate the human albumin/IR780/iron oxide nanocomplexes with a hydrophobic core and a hydrophilic shell consisting of human serum albumin and polyethylene glycol (Lin et al. 2018). Focused ultrasound studies under the control of MRI were followed by intravenous infusion of an adeno-associated virus serotype vector expressing green fluorescent protein as a marker (Stavarache et al. 2018). In work by Rolla and coworkers the Mn^{2+}-complexes of six original amphiphilic ligands embodying one or two aliphatic chains were evaluated as potential magnetic resonance imaging contrast agents (Rolla et al. 2018).

Table 1. Advances of the use of MRI and MRS

Methods	Aim / Results	References
MRI	To develop a method for improved cell uptake of ferucarbotran magnetic nanoparticles contained in Resovist by modifying the nanoparticle shell with human serum albumin	Aurich et al. 2017
	To develop O_2-generating HSA-MnO_2-Ce6 nanoparticles to overcome tumor hypoxia and thus enhance the photodynamic effect for bladder cancer therapy	Lin et al. 2018
	To develop multifunctional albumin/superparamagnetic iron oxide nanoparticle system to deliver a photothermal agent	Lin et al. 2018
	To develop nanoparticles consistet of gadolinium/1,4,7,10-tetraazacyclododecane-1,4,7-tetracetic acid complex with 6,6-phenyl-C61 butyric acid and upon further modification with HSA	Zhang et al. 2016
	To design innovative HAS based nanoparticles loaded with silencing RNA and grafted with gadolinium complexes	Mertz et al. 2014
	To develop nanoparticles conjugated HAS with gadolinium diethylenetriaminepentaacetic acid using carbodiimide chemistry	Watcharin et al. 2014
	To develop serum albumin targeted MRI probes based on a single amino acid Gd complex	Boros and Caravan 2013
^{19}F NMR	To develop interaction of capecitabine and gefitinib with human serum albumin. Interaction with human serum albumin caused ^{19}F NMR signal shifts	Wu et al. 2016

A macromolecular magnetic resonance imaging contrast agent was successfully synthesized by conjugating gadolinium/1,4,7,10-tetraaza-cyclododecane-1,4,7-tetracetic acid complex with 6,6-phenyl-C61 butyric acid and upon further modification with human serum albumin (Zhang et al.

2016). The binding capacities to human serum albumin of two anticancer drugs, capecitabine and gefitinib, were compared via an approach combining ^{19}F NMR, ^{1}H saturation transfer difference NMR, circular dichroism and docking simulations (Wu et al. 2016). Research done by Haubner and coworkers showed that [^{68}Ga]NOTA-GSA and [^{68}Ga]DTPA-GSA can be used equivalent for MRI imaging of hepatic function together with positron emission tomography (Haubner et al. 2017). It is shown that high affinity Human serum albumin binding is presented in tumors lowers that the binding potential of radiotracers (Digilio et al. 2014). Human serum albumin-based nanoparticles loaded with silencing RNA and grafted with gadolinium complexes having average sizes ranging from ca. 50 to 150 nm according to the siRNA/HSA composition (Mertz et al. 2014). Human serum albumin nanoparticles appear to be a suitable carrier due to their safety and feasibility of functionalization. In the present study HSA nanoparticles were conjugated with gadolinium diethylenetriaminepentaacetic acid using carbodiimide chemistry (Watcharin et al. 2014). The Gd(III) complex of DO3A-N-α-aminopropionate, Gd(DOTAla), was used to generate a small library of putative MRI probes targeted to human serum albumin (Boros et al. 2013). Magnetic resonance was used to investigate lymphangiography in mice and primates with intradermal Gadofosveset and human serum albumin (Nakajima et al. 2014). Polymeric nanoparticles prepared from human serum albumin, loaded with gadolinium-diethylenetriaminepentaacetic acid (Gd-HSA-NP), and coated with folic acid were imaged by magnetic resonance imaging (Korkusuz et al. 2012). Some gadolinium complexes, commonly used as MRI contrast agents, have a high affinity for human serum albumin which enhances their efficacy (Henoumont et al. 2012; Moriggi et al. 2012). Iron oxide nanoparticles SPIO-MRI may be a helpful, non-invasive method for the evaluation of hepatic functional reserve, and this study suggests that Kupffer cell function is closely correlated with hepatocyte function in patients with chronic viral hepatitis (Tonan et al. 2011). Gd-C(4)-thyroxin-DTPA with human serum albumin was synthesized as a potential contrast agent (Henoumont et al. 2010). Immunohistochemical analyses revealed that the Gd-Rho-human serum albumin was localized to macrophages by MRI (Watcharin et al. 2015). The nanoparticles contained dysprosium(iii) ion with human serum albumin were assembled into mono-disperse micelles and the magnetic and optical properties of the micelles were examined (Harris et al. 2016). The purpose of the study performed by Matoba and coworkers was to compare the assessment of functional severity on in vivo hepatic ^{31}P-MRS (Matoba et al. 2000). It was analysed whether clinical variables are useful for predicting mucosal healing in various types of nanoparticels containing human serum albumin (Bamba et al. 2014). Also, serum fentanyl with human serum albumin was measured by using MR between cancer pain patients (Barratt et al. 2014). Magnetic resonance spectroscopy measurements of the lactate methyl proton in rat brain C6 glioma tissue was acquired in the presence of an off-resonance irradiation field (Luo et al. 1999).

CONCLUSION

In reviewed studies Magnetic resonance imaging techniques were used for non-invasively mapping drug distribution and drug conjugated with human serum albumin, T_1 and T_2 relaxation times and the apparent diffusion coefficient in conjunction with a drug.

ACKNOWLEDGMENTS

Dorota Bartusik-Aebisher acknowledges support from the National Center of Science NCN (New drug delivery systems-MRI study, Grant OPUS-13 number 2017/25/B/ST4/02481).

REFERENCES

Aurich, K., Wesche, J., Palankar, R., Schlüter, R., Bakchoul, T., Greinacher, A. (2017). Magnetic Nanoparticle Labeling of Human Platelets from

Platelet Concentrates for Recovery and Survival Studies. *ACS applied materials & interfaces*, 9:34666 - 34673.

Bamba, S., Tsujikawa, T., Ban, H., Imaeda, H., Inatomi, O., Nishida, A., Sasaki, M., Andoh, A., Fujiyama, Y. (2014). Predicting Mucosal Healing in Crohn's Disease Using Practical Clinical Indices with Regard to the Location of Active Disease. *Hepatogastroenterology*, 61(131):689 - 96.

Barratt, D. T., Bandak, B., Klepstad, P., Dale, O., Kaasa, S., Christrup, L. L., Tuke, J., Somogyi, A. A. (2014). Genetic, pathological and physiological determinants of transdermal fentanyl pharmacokinetics in 620 cancer patients of the EPOS study. *Pharmacogenetics and genomics*, 24(4):185 - 94.

Boros, E., Caravan, P. (2013). Structure-relaxivity relationships of serum albumin targeted MRI probes based on a single amino acid Gd complex. *Journal of medicinal chemistry*, 56:1782 - 6.

Digilio, G., Tuccinardi, T., Casalini, F., Cassino, C., Dias, D. M., Geraldes, C. F., Catanzaro, V., Maiocchi, A., Rossello, A. (2014). Study of the binding interaction between fluorinated matrix metalloproteinase inhibitors and Human Serum Albumin. *European journal of medicinal chemistry*, 79:13 - 23.

Erdal, E., Demirbilek, M., Yeh, Y., Akbal, Ö., Ruff, L., Bozkurt, D., Cabuk, A., Senel, Y., Gumuskaya, B., Algın, O., Colak, S., Esener, S., Denkbas, E. B. (2018). A Comparative Study of Receptor-Targeted Magnetosome and HSA-Coated Iron Oxide Nanoparticles as MRI Contrast-Enhancing Agent in Animal Cancer Model. *Applied biochemistry and biotechnology*, 185:91 - 113.

Harris, M., Vander, Elst L., Laurent, S., Parac-Vogt, T. N. (2016). Magnetofluorescent micelles incorporating Dy(III)-DOTA as potential bimodal agents for optical and high field magnetic resonance imaging. *Dalton transactions: an international journal of inorganic chemistry/ RSoC*, 45(11):4791 - 801.

Henoumont, C., Vander, Elst L., Laurent, S., Muller, R. N. (2010). Synthesis and physicochemical characterization of Gd-C4-thyroxin-DTPA, a potential MRI contrast agent. Evaluation of its affinity for human serum albumin by proton relaxometry, NMR diffusometry, and electrospray mass spectrometry. *The journal of physical chemistry. B*, 114(10):3689 - 97.

Henoumont, C., Laurent, S., Muller, R. N., Vander, Elst L. (2012). Effect of nonenzymatic glycosylation on the magnetic resonance imaging (MRI) contrast agent binding to human serum albumin. *Journal of medicinal chemistry*, 55(8):4015 - 9.

Korkusuz, H., Ulbrich, K., Bihrer, V., Welzel, K., Chernikov, V., Knobloch, T., Petersen, S., Huebner, F., Ackermann, H., Gelperina, S., Korkusuz, Y., Kromen, W., Hammerstingl, R., Haupenthal, J., Fiehler, J., Zeuzem, S., Kreuter, J., Vogl, T. J., Piiper, A. (2012). Contrast enhancement of the brain by folate-conjugated gadolinium-diethylenetriaminepentaacetic acid-human serum albumin nanoparticles by magnetic resonance imaging. *Molecular imaging*, 11(4):272 - 9.

Lenora, C. U., Carniato, F., Shen, Y., Latif, Z., Haacke, E. M., Martin, P. D., Botta, M., Allen, M. J. (2017). Structural Features of Europium(II)-Containing Cryptates That Influence Relaxivity. *Chemistry*, 23:15404-15414.

Lin, T., Zhao, X., Zhao, S., Yu, H., Cao, W., Chen, W., Wei, H., Guo, H. (2018). O2-generating MnO2 nanoparticles for enhanced photodynamic therapy of bladder cancer by ameliorating hypoxia. *Theranostics*, 8:990 - 1004.

Lin, S. Y., Huang, R. Y., Liao, W. C., Chuang, C. C., Chang, C. W. (2018). Multifunctional PEGylated Albumin/IR780/Iron Oxide Nanocomplexes for Cancer Photothermal Therapy and MR Imaging. *Nanotheranostics*, 2:106 - 116.

Luo, Y., Rydzewski, J., de Graaf, R. A., Gruetter, R., Garwood, M., Schleich, T. (1999). In vivo observation of lactate methyl proton magnetization transfer in rat C6 glioma. *Magnetic resonance in medicine*, 41(4):676 - 85.

Matoba, M., Tonami, H., Yokota, H., Higashi, K., Yamamoto, I. (2000). Assessment of functional severity on in vivo hepatic 31P-MRS in diffuse hepatic disease: comparative studies with 99mTc-GSA. *Nihon acta radiologica*, 60(8):439 - 44.

Mertz, D., Affolter-Zbaraszczuk, C., Barthès, J., Cui, J., Caruso, F., Baumert, T. F., Voegel, J. C., Ogier, J., Meyer, F. (2014). Templated assembly of albumin-based nanoparticles for simultaneous gene silencing and magnetic resonance imaging. *Nanoscale*, 6:11676 - 80.

Morine, Y., Enkhbold, C., Imura, S., Ikemoto, T., Iwahashi, S., Saito, Y. U., Yamada, S., Utsunomiya, T., Shimada, M. (2017). Accurate Estimation of Functional Liver Volume Using Gd-EOB-DTPA MRI Compared to MDCT/99mTc-SPECT Fusion Imaging. *Anticancer research*, 37:5693 - 5700.

Moriggi, L., Yaseen, M. A., Helm, L., Caravan, P. (2012). Serum albumin targeted, pH-dependent magnetic resonance relaxation agents. *Chemistry*, 18(12):3675 - 86.

Nakajima, T., Turkbey, B., Sano, K., Sato, K., Bernardo, M., Hoyt, R. F., Choyke, P. L., Kobayashi, H. (2014). MR lymphangiography with intradermal gadofosveset and human serum albumin in mice and primates. *Journal of magnetic resonance imaging: JMRI*, 40(3): 691 - 7.

Nasr, S. H., Kouyoumdjian, H., Mallett, C., Ramadan, S., Zhu, D. C., Shapiro, E. M., Huang, X. (2018). Detection of β-Amyloid by Sialic Acid Coated Bovine Serum Albumin Magnetic Nanoparticles in a Mouse Model of Alzheimer's Disease. *Small*, 14.

Rolla, G., De Biasio, V., Giovenzana, G. B., Botta, M., Tei, L. (2018). Supramolecular assemblies based on amphiphilic Mn^{2+}-complexes as high relaxivity MRI probes. *Dalton transactions: an international journal of inorganic chemistry/RSoC*, doi: 10.1039/c8dt01250d.

Shamsutdinova, N., Zairov, R., Nizameev, I., Gubaidullin, A., Mukhametshina, A., Podyachev, S., Ismayev, I., Kadirov, M., Voloshina, A., Mukhametzyanov, T., Mustafina, A. (2018). Tuning magnetic relaxation properties of "hard cores" in core-shell colloids by modification of "soft shell". *Colloids and surfaces. B, Biointerfaces*, 162:52 - 59.

Stavarache, M. A., Petersen, N., Jurgens, E. M., Milstein, E. R., Rosenfeld, Z. B., Ballon, D. J., Kaplitt, M. G. (2018). Safe and stable noninvasive focal gene delivery to the mammalian brain following focused ultrasound. *Journal of neurosurgery*, 27:1 - 10.

Tonan, T., Fujimoto, K., Qayyum, A., Azuma, S., Ishibashi, M., Ueno, T., Ono, N., Akiyoshi, J., Matsushita, S., Hayabuchi, N., Kawaguchi, T., Sata, M. (2011). Correlation of Kupffer cell function and hepatocyte function in chronic viral hepatitis evaluated with superparamagnetic iron oxide-enhanced magnetic resonance imaging and scintigraphy using technetium-99m-labelled galactosyl human serum albumin. *Experimental and therapeutic medicine*, 2(4):607 - 613.

Watcharin, W., Schmithals, C., Pleli, T., Köberle, V., Korkusuz, H., Huebner, F., Zeuzem, S., Korf, H. W., Vogl, T. J., Rittmeyer, C., Terfort, A., Piiper, A., Gelperina, S., Kreuter, J. (2014). Biodegradable human serum albumin nanoparticles as contrast agents for the detection of hepatocellular carcinoma by magnetic resonance imaging. *European Journal of Pharmaceutics and Biopharmaceutics*, 87:132 - 41.

Watcharin, W., Schmithals, C., Pleli, T., Köberle, V., Korkusuz, H., Hübner, F., Waidmann, O., Zeuzem, S., Korf, H. W., Terfort, A., Gelperina, S., Vogl, T. J., Kreuter, J., Piiper, A. (2015). Detection of hepatocellular carcinoma in transgenic mice by Gd-DTPA- and rhodamine 123-conjugated human serum albumin nanoparticles in T1 magnetic resonance imaging. *Journal of controlled release*, 199:63 - 71.

Wu, D., Yan, J., Sun, P., Xu, K., Li, S., Yang, H., Li, H. (2016). Comparative analysis of the interaction of capecitabine and gefitinib with human serum albumin using ^{19}F nuclear magnetic resonance-based approach. *Journal of pharmaceutical and biomedical analysis*, 129:15 - 20.

Zhang, Y., Zou, T., Guan, M., Zhen, M., Chen, D., Guan, X., Han, H., Wang, C., Shu, C. (2016). Synergistic Effect of Human Serum Albumin and Fullerene on Gd-DO3A for Tumor-Targeting Imaging. *ACS applied materials & interfaces*, 8:11246 - 54.

In: Human Serum Albumin
Editor: Dianne Cohen

ISBN: 978-1-53614-787-2
© 2019 Nova Science Publishers, Inc.

Chapter 5

THE INFLUENCE OF FLUORINE-19 DRUG LABELING ON BINDING TO HUMAN SERUM ALBUMIN

Dorota Bartusik-Aebisher*, David Aebisher and Zuzanna Bober*
Faculty of Medicine, University of Rzeszow, Rzeszow, Poland

ABSTRACT

We review the major approaches for the characterization of Human Serum Albumin as a fluorinated drug delivery agent and fluorinated albumin influence on drug binding. Synthesis and characterization of fluorinated conjugates of albumin, adsorbed Human Serum Albumin on surfaces containing CF_3 are also discussed. The databases such as PubMed, ScienceDirect and Springer were utilized to search the literature for relevant articles.

Keywords: Human serum albumin, drug activity, ^{19}F NMR

---

* Corresponding Author Email: dbartusik-aebisher@ur.edu.pl.

^{19}F NMR techniques were employed to characterize the binding property of the widely used general anesthetic with human serum albumin (Shikii et al. 2004). It was found that ^{19}F(^{1}H) NOE and 2D ^{1}H-^{19}F HOESY experiments detected intermolecular NOEs between halothane ^{19}F and HSA protons (Shikii et al. 2004). Human serum albumin can significantly enhance the solubility of some compounds e.g., AM5206 in aqueous environment (Zhuang et al. 2013). This findings also help in the development of suitable formulations of the lipophilic AM5206 and its congeners for their effective delivery to specific target sites in the brain (Zhuang et al. 2013). Molecular docking simulation was used to establish a molecular binding modelby using NMR experiments (Yan et al. 2016; del Prado et al. 2010). Identification of compounds from chemical libraries that bind to macromolecules by use of NMR spectroscopy has gained increasing importance during recent years. ^{19}F NMR spectroscopy was used for the screening of ligands that bind to proteins, which also provides qualitative information about relative binding strengths and the presence of multiple binding sites (Tengel et al. 2004). It was found that library members which are bound to a target protein could be identified directly from line broadening and/or induced chemical shifts in a single, one-dimensional ^{19}F NMR spectrum (Tengel et al. 2004). Human serum albumin has antioxidant capacity (Aime et al. 1999; Köster et al. 1996). Human serum albumin binding was also investigated by difference ^{1}H NMR and by measuring the increase in the ^{19}F NMR signal (Bertucci et al. 1995). Study of ligand-macromolecular interactions by ^{19}F nuclear magnetic resonance spectroscopy represents molecular biochemical and pharmaceutical information. ^{19}F NMR was used to study of these interactions in vivo, as well for receptor binding and metabolic tracing of fluorinated drugs and proteins are discussed (Jenkins 1991).

Binding and co-binding of various 19F-labeled ligands to human serum albumin has been studied using ^{19}F NMR (Jenkins 1990). This technique proves to be a powerful methodology for studying ligand and drug binding to HSA that is free from some of the pitfalls associated with more traditional techniques such as equilibrium dialysis (Jenkins 1990).

Bovine serum albumin was derivatized, under mild conditions, with trifluoromethyl-dinitrophenyl, a haptenic group cross-reacting with trinitrophenyl (Bischoff et al. 1984). Immune responses were studied in adult and young mice exposed to pneumococcal 6A and 19F polysaccharides (Lin et al. 1982). Immunization of mice with 19F polysaccharide-protein conjugates resulted in formation of more antibody than was found in the control group (Lee et al. 2001; Concepcion and Frasch 1998; Lee and Lin 1991). The ^{19}F NMR spectrum of triflupromazine hydrochloride in a buffer solution showed a single sharp signal of the TFZ CF3 group at 13.5 ppm from the external trifluoroacetic acid (Kitamura et al. 2004). The Ru(III) complexes indazolium [trans-RuCl4(1H-indazole)2] (KP1019) and sodium [trans-RuCl4(1H-indazole)2] (NKP-1339) are leading candidates for the next generation of metal-based chemotherapeutics (Kitamura et al. 2016). Cyclic tetrapeptides containing a β2,2-amino acid with either two 2-naphthyl-methylene or two para-CF3-benzyl side chains, along with their interaction with the main plasma protein human serum albumin was sudied (Sivertsen et al. 2014). It is published that Sevoflurane (CH2F-O-CH[CF3]2) reacts with carbon dioxide absorbents to produce Compound A (CH2F-O-C[=CF2][CF3]) (Eger et al. 1997). Knowledge of the solubility of Compound A, CH2F-O-C(=CF2)(CF3), in blood and other solvents would aid in the definition of its kinetics. Accordingly, we determined solvent/gas partition coefficients of Compound A for saline (0.166 +/- 0.002 [mean +/- SD; n = 4]) and olive oil (20.1 +/- 1.1 [n = 4]) (Eger et al. 1996). The strong albumin binding to fluorocarbon surfaces may be exploited clinically to enhance the retention of albumin (Bohnert et al. 1990). It is shown that high affinity albumin binding constitutes a theoretical limitation for the specificity achievable by MMP-inhibitors because albumin accumulating aspecifically in tumours lowers the binding potential of radiotracers (Digilio et al. 2014). Using fluorine labeling of the drug and 19F NMR, we determined Human Serum Albumin affinity for liraglutide in two glycated albumin models. Because diabetes is a progressive disease, the effect of glycated albumin on liraglutide affinity found here is important to consider when diabetes is managed with this drug (Gajahi et al. 2018). ^{19}F NMR spectroscopy was usedfor for the detection of the starting and enzymatically modified substrates (Dalvit et al. 2003). The conjugate allows for direct optical and ^{19}F magnetic resonance cancer imaging and release of the drug upon addition of glutathione (Lisitskiy 2017). Human serum albumin binding to two anticancer drugs, capecitabine and gefitinib, were compared via an approach combining ^{19F} NMR (Wu et al. 2016). By comparing the properties of N-perfluorotoluene-homocystamide of albumin with N-homocysteinylated albumin, it has been revealed that blocking of the alpha-amino group of the homocysteine residue in the fluorinated albumin conjugate inhibits the dangerous aggregation process, as well as free radical formation (Chubarov et al. 2015). ^{19}F NMR signals have been observed in human serum for free and plasma-protein bound 5'-deoxy-5-fluorouridine (Meynial et al. 1988).

Uridine analogs 5'dFUrd are transported in the blood by transporter proteins like human serum albumin (Ishtikhar et al. 2014; Kumar et al. 2005 Di Stefano et al. 2002). A fluorinated phthalocyanine and its non-fluorinated analogue were selected to evaluate the potential enhancement of fluorination on photophysical, photochemical and redox properties as well as on biological activity in cellular and animal models (Pucelik et al. 2016). Derivatives of 2'-deoxyuridine and of the anticancer agent 5-fluoro-2'-deoxyuridine were linked indirectly via a human serum albumin carrier to the murine antiosteosarcoma monoclonal antibody 791T/36 (Henn et al. 1993).

A N-trifluoroacetyl-protected amino acid containing a thioester function, 2,2,2-trifluoro-N-(2-oxo-tetrahydrothiophen-3-yl)acetamide was used to prepare a fluorine-labeled N-homocysteinylated protein and was characterized by ^{19}F NMR spectroscopy (Chubarov et al. 2011). Complexation between human serum albumin and two different surfactants, one fully fluorinated sodium perfluorooctanoate was studied using zeta-potential measurements and difference spectroscopy (Sabín et al. 2006).

In the present study, 3-(fluorobenzylideneamino)-6-chloro-1-(3,3-dimethylbutanoyl)-phenyl-2,3-dihydroquinazolin-4(1H)-one derivatives have been designed and synthesized to study the interaction between fluorine substituted dihydroquinazoline derivatives with human serum albumin using fluorescence, circular dichroism and Fourier transform infrared spectroscopy (Wang et al. 2016).

Table 1. Examples of literature on fluorinated drugs and detection methods

Aim	Methods	Results	References
Binding and co-binding of various ^{19}F-labeled ligands to human serum albumin	^{19}F NMR	5-F-L-tryptophan and 5-F-salicylic acid are capable of binding independently to two sites on HSA at the same time, reveals allosteric interactions between 5-F-L-Trp and warfarin co-bound to human serum albumin	Jenkins and Lauffer 1990
5-Fluorouracil is carried in the serum by plasma proteins	UV, CD, ^{1}H and ^{19}F NMR	Binding was investigated by difference ^{1}H NMR and by measuring the increase in the ^{19}F NMR signal	Bertucci et al. 1995
The antioxidant capacity of a given compound has been scaled in terms of an adimensional parameter, kF, that represents the ratio between the scavenger abilities of the fluorinated detector and the competitor	^{19}F-NMR spectroscopy	The kF value for serum albumin is much larger than that predicted from the reported k2OH value.	Aime et al. 1999
sensitive NMR method for rapid, efficient, and reliable biochemical screening is presented	^{19}F NMR spectroscopy	The method named 3-FABS (three fluorine atoms for biochemical screening) requires the labeling of the substrate with a CF_3 moiet. The method allows for high-quality screening of large compound or natural product extract collections and for measuring their IC(50) values.	Dalvit et al. 2003

Table 1. (Continued)

Aim	Methods	Results	References
Observe general anesthetic halothane to human serum albumin	¹⁹F NMR	indicating the interaction of halothane with human serum albumin	Shikii et al. 2004
fluorinated compounds was assembled and investigated for binding to the two bacterial chaperones PapD and FimC, and also to human serum albumin	¹⁹F NMR spectroscopy	target protein could be identified directly from line broadening and/or induced chemical shifts in a single, one-dimensional ¹⁹F NMR spectrum	Tengel et al. 2004
binding of triflupromazine to bovine and human serum albumins	¹⁹F NMR	¹⁹F NMR is very useful for obtaining important detailed information regarding the binding of fluorinated drugs to serum albumins	Kitamura et al. 2004
interactions of AM5206 with a representative AEA carrier protein	¹⁹F-NMR	AM5206 primarily binds to two distinct sites within HSA	Zhuang et al. 2013
interaction of arylsulfone-based inhibitors of Matrix Metalloproteinases with Human Serum Albumin	¹⁹F NMR	high affinity albumin binding constitutes a theoretical limitation for the specificity achievable by MMP-inhibitors as MMP-targeted PET tracers in cancer imaging, because albumin accumulating aspecifically in tumours lowers the binding potential of radiotracers	Digilio et al. 2014
develop molecular probes with appropriate NMR characteristics	¹⁹F NMR,	fluorine-labeled homocysteine thiolactone has been employed	Chubarov et al. 2015
biocompatibility for in vivo applications using ¹⁹F MRI		perfluorotoluene group as a ¹⁹F-containing tag into human serum albumin, perfluoro-toluene-labeled albumin has been demonstrated to act as a promising agent for in vivo ¹⁹F MRI	

Aim	Methods	Results	References
mechanisms of tyrosine-kinase inhibitor nilotinib to human serum albumin	^{19}F NMR	NIL mainly bound to the hydrophobic cavity of HSA's subdomain IIA	Yan et al. 2016
CF$_3$ Derivatives of the Anticancer Ru(III) Complexes KP1019, NKP-1339, and Their Imidazole and Pyridine Analogues	^{19}F NMR	Addition of CF$_3$ groups also provided a spectroscopic handle for ^{19}F NMR studies of ligand exchange processes and protein interactions. Addition of CF$_3$ groups enhances the activity of the indazole complex against A549 nonsmall cell lung carcinoma cells	Chang et al. 2016
interaction of capecitabine and gefitinib with human serum albumin	^{19}F NMR	results showed that the two drugs interaction with human serum albumin caused ^{19}F NMR signal shifted to different directions, Capecitabine had accurate binding site and higher binding affinity than gefitinib.	Wu et al. 2016
synthesis and properties of a new multimodal theranostic conjugate based on an anticancer fluorinated nucleotide conjugated with a dual-labeled albumin	^{19}F NMR ^{19}F MRI	fluorine-labeled homocysteine thiolactone has been used as functional handle to synthesize the fluorinated albumin and couple it with a chemotherapeutic agent 5-trifluoromethyl-2'-deoxyuridine 5'-monophosphate conjugate allows for direct optical and ^{19}F magnetic resonance cancer imaging and release of the drug upon addition of glutathione.	Lisitskiy et al. 2017

CONCLUSION

^{19}F NMR spectroscopy should be a valuable addition to existing NMR techniques.

ACKNOWLEDGMENTS

Dorota Bartusik-Aebisher acknowledges support from the National Center of Science NCN (New drug delivery systems-MRI study, Grant OPUS-13 number 2017/25/B/ST4/02481).

REFERENCES

Aime, S., Calzoni, S., Digilio, G., Giraudo, S., Fasano, M., Maffeo, D. (1999). A novel 19F-NMR method for the investigation of the antioxidant capacity of biomolecules and biofluids. *Free Radical Biology & Medicine*, 27:356 - 63.

Bertucci, C., Ascoli, G., Uccello-Barretta, G., Di Bari, L., Salvadori, P. (1995). The binding of 5-fluorouracil to native and modified human serum albumin: UV, CD, and 1H and 19F NMR investigation. *Journal of Pharmaceutical and Biomedical Analysis*, 13:1087 - 93.

Bischoff, P., Maugras, M., Poignant, S., Oth, D. (1984). Cell surface modifications with trifluoromethyl dinitrophenyl-soluble protein conjugates: immunogenic role of noncovalently bound hapten. *International Archives of Allergy and Applied Immunology*, 75:20 - 6.

Bohnert, J. L., Fowler, B. C., Horbett, T. A., Hoffman, A. S. (1990). Plasma gas discharge deposited fluorocarbon polymers exhibit reduced elutability of adsorbed albumin and fibrinogen. *Journal of Biomaterials Science. Polymer Edition*, 1:279 - 97.

Chang, S. W., Lewis, A. R., Prosser, K. E., Thompson, J. R., Gladkikh, M., Bally, M. B., Warren, J. J., Walsby, C. J. (2016). CF3 Derivatives of the Anticancer Ru(III) Complexes KP1019, NKP-1339, and Their

Imidazole and Pyridine Analogues Show Enhanced Lipophilicity, Albumin Interactions, and Cytotoxicity. *Inorganic Chemistry*, 55:4850 - 63.

Chubarov, A. S., Zakharova, O. D., Koval, O. A., Romaschenko, A. V., Akulov, A. E., Zavjalov, E. L., Razumov, I. A., Koptyug, I. V., Knorre, D. G., Godovikova, T. S. (2015). Design of protein homocystamides with enhanced tumor uptake properties for (19)F magnetic resonance imaging. *Bioorganic & Medicinal Chemistry*, 23:6943 - 54.

Chubarov, A. S., Shakirov, M. M., Koptyug, I. V., Sagdeev, R. Z., Knorre, D. G., Godovikova, T. S. (2011). Synthesis and characterization of fluorinated homocysteine derivatives as potential molecular probes for ^{19}F magnetic resonance spectroscopy and imaging. *Bioorganic & Medicinal Chemistry Letters*, 21:4050 - 3.

Concepcion, N., Frasch, C. E. (1998). Evaluation of previously assigned antibody concentrations in pneumococcal polysaccharide reference serum 89SF by the method of cross-standardization. *Clinical and Diagnostic Laboratory Immunology*, 5:199 - 204.

Dalvit, C., Ardini, E., Flocco, M., Fogliatto, G. P., Mongelli, N., Veronesi, M. (2003). A general NMR method for rapid, efficient, and reliable biochemical screening. *Journal of the American Chemical Society*, 125:14620 - 5.

Del Prado, G., Ruiz, V., Naves, P., Rodríguez-Cerrato, V., Soriano, F., del Carmen Ponte, M. (2010). Biofilm formation by Streptococcus pneumoniae strains and effects of human serum albumin, ibuprofen, N-acetyl-l-cysteine, amoxicillin, erythromycin, and levofloxacin. *Diagnostic Microbiology and Infectious Disease*, 67:311 - 8.

Digilio, G., Tuccinardi, T., Casalini, F., Cassino, C., Dias, D. M., Geraldes, C. F., Catanzaro, V., Maiocchi, A., Rossello, A. (2014). Study of the binding interaction between fluorinated matrix metalloproteinase inhibitors and Human Serum Albumin. *European Journal of Medicinal Chemistry*, 79:13 - 23.

Di Stefano, G., Lanza, M., Busi, C., Barbieri, L., Fiume, L. (2002). Conjugates of nucleoside analogs with lactosaminated human albumin to selectively increase the drug levels in liver blood: requirements for a regional chemotherapy. *The Journal of Pharmacology and Experimental Therapeutics*, *301*:638 - 42.

Eger, E. I. 2nd1, Gong, D., Koblin, D. D., Bowland, T., Ionescu, P., Laster, M. J., Weiskopf, R. B. (1997). Dose-related biochemical markers of renal injury after sevoflurane versus desflurane anesthesia in volunteers. *Anesthesia and Analgesia*, *85*:1154 - 63.

Eger, E. I. 2nd1, Ionescu, P., Koblin, D. D., Weiskopf, R. B. (1996). Compound A: solubility in saline and olive oil; destruction by blood. *Anesthesia and Analgesia*, *83*:849 - 53.

Gajahi Soudahome, A., Catan, A., Giraud, P., Assouan Kouao, S., Guerin-Dubourg, A., Debussche, X., Le Moullec, N., Bourdon, E., Bravo, S. B., Paradela-Dobarro, B., Álvarez, E., Meilhac, O., Rondeau, P., Couprie, J. (2018). Glycation of human serum albumin impairs binding to the glucagon-like peptide-1 analogue liraglutide. *The Journal of Biological Chemistry*, *293*:4778 - 4791.

Henn, T. F., Garnett, M. C., Chhabra, S. R., Bycroft, B. W., Baldwin, R. W. (1993). Synthesis of 2'-deoxyuridine and 5-fluoro-2'-deoxyuridine derivatives and evaluation in antibody targeting studies. *Journal of Medicinal Chemistry*, *36*:1570 - 9.

Ishtikhar, M., Rabbani, G., Khan, R. H. (2014). Interaction of 5-fluoro-5'-deoxyuridine with human serum albumin under physiological and non-physiological condition: a biophysical investigation. *Colloids and Surfaces. B, Biointerfaces*, *123*:469 - 77.

Jenkins, B. G. (1991). Detection of site-specific binding and co-binding of ligands to macromolecules using 19F NMR. *Life Sciences*, 48:1227 - 40.

Jenkins, B. G., Lauffer, R. B. (1990). Detection of site-specific binding and co-binding of ligands to human serum albumin using 19F NMR. *Molecular Pharmacology*, *37*:111 - 8.

Kitamura, K., Kume, M., Yamamoto, M., Takegami, S., Kitade, T. (2004). 19F NMR spectroscopic study on the binding of triflupromazine to bovine and human serum albumins. *Journal of Pharmaceutical and Biomedical Analysis*, *36*:411 - 4.

Kumar, Y., Muzammil, S., Tayyab, S. (2005). Influence of fluoro, chloro and alkyl alcohols on the folding pathway of human serum albumin. *Journal of Biochemistry*, *138*:335 - 41.

Köster, U., Mayer, D., Deger, H. M., DeKant, W. (1996). Biotransformation of the aerosol propellant 1,1,1,2,3,3,3-heptafluoropropane (HFA-227): lack of protein binding of the metabolite hexafluoroacetone. *Drug Metabolism and Disposition: the Biological Fate of Chemicals*, *24*:906 - 10.

Lee, C. J., Lin, K. T. (1981). Studies on vaccine control and immunogenicity of polysaccharides of Streptococcus pneumoniae. *Reviews of Infectious Diseases*, *3*:51 - 60.

Lee, C. J., Wang, T. R., Frasch, C. E. (2001). Immunogenicity in mice of pneumococcal glycoconjugate vaccines using pneumococcal protein carriers. *Vaccine*, *19*:3216 - 25.

Lin, K. T., Lee, C. J. (1982). Immune response of neonates to pneumococcal polysaccharide-protein conjugate. *Immunology*, 46:333 - 42.

Lisitskiy, V. A., Khan, H., Popova, T. V., Chubarov, A. S., Zakharova, O. D., Akulov, A. E., Shevelev, O. B., Zavjalov, E. L., Koptyug, I. V., Moshkin, M. P., Silnikov, V. N., Ahmad, S., Godovikova, T. S. (2017). Multifunctional human serum albumin-therapeutic nucleotide conjugate with redox and pH-sensitive drug release mechanism for cancer theranostics. *Bioorganic & Medicinal Chemistry Letters*, *27*:3925 - 3930.

Meynial, D., Lopez, A., Malet-Martino, M. C., Hoffmann, J. S., Martino, R. (1988). Application of fluorine-19 nuclear magnetic resonance to the determination of plasma-protein binding of 5'-deoxy-5-fluorouridine, a new antineoplastic fluoropyrimidine. *Journal of Pharmaceutical and Biomedical Analysis*, *6*:47 - 59.

Pucelik, B., Gürol, I., Ahsen, V., Dumoulin, F., Dąbrowski, J. M. (2016). Fluorination of phthalocyanine substituents: Improved photoproperties and enhanced photodynamic efficacy after optimal micellar formulations. *European Journal of Medicinal Chemistry*, *124*:284 - 298.

Sivertsen, A., Tørfoss, V., Isaksson, J., Ausbacher, D., Anderssen, T., Brandsdal, B. O., Havelkova, M., Skjørholm, A. E., Strøm, M. B. (2014). Anticancer potency of small linear and cyclic tetrapeptides and pharmacokinetic investigations of peptide binding to human serum albumin. *Journal of Peptide Science*, 20:279 - 91.

Shikii, K., Sakurai, S., Utsumi, H., Seki, H., Tashiro, M. (2004). Application of the 19F NMR technique to observe binding of the general anesthetic halothane to human serum albumin. *Analytical Sciences*, 20:1475 - 7.

Tengel, T., Fex, T., Emtenas, H., Almqvist, F., Sethson, I., Kihlberg, J. (2004). Use of 19F NMR spectroscopy to screen chemical libraries for ligands that bind to proteins. *Organic & Biomolecular Chemistry*, 2:725 - 31.

Wang, Y., Zhu, M., Liu, F., Wu, X., Pan, D., Liu, J., Fan, S., Wang, Z., Tang, J., Na, R., Li, Q. X., Hua, R., Liu, S. (2016). Comparative Studies of Interactions between Fluorodihydroquinazolin Derivatives and Human Serum Albumin with Fluorescence Spectroscopy. *Molecules, 21*.

Wu, D., Yan, J., Sun, P., Xu, K., Li, S., Yang, H., Li, H. (2016). Comparative analysis of the interaction of capecitabine and gefitinib with human serum albumin using (19)F nuclear magnetic resonance-based approach. *Journal of Pharmaceutical and Biomedical Analysis, 129*:15 - 20.

Yan, J., Wu, D., Sun, P., Ma, X., Wang, L., Li, S., Xu, K., Li, H. (2016). Binding mechanism of the tyrosine-kinase inhibitor nilotinib to human serum albumin determined by 1H STD NMR, 19F NMR, and molecular modeling. *Journal of Pharmaceutical and Biomedical Analysis, 124*:1 - 9.

Zhuang, J., Yang, D. P., Tian, X., Nikas, S. P., Sharma, R., Guo, J. J., Makriyannis, A. (2013). Targeting the Endocannabinoid System for Neuroprotection: A 19F-NMR Study of a Selective FAAH Inhibitor Binding with an Anandamide Carrier Protein, HSA. *Journal of Pharmaceutics & Pharmacology, 1:*002.

In: Human Serum Albumin
Editor: Dianne Cohen

ISBN: 978-1-53614-787-2
© 2019 Nova Science Publishers, Inc.

Chapter 6

TYROSINE DETECTION IN MICROALBUMINURIA

Dorota Bartusik-Aebisher[1,*]*, David Aebisher*[1]*,
Sabina Galiniak*[1]*, Łukasz Ożóg*[1]
and Małgorzata Marć[1]

[1]Faculty of Medicine, University of Rzeszów, Rzeszów, Poland

ABSTRACT

The assessment of subtle changes in albumin concentration, the primary macromolecular component of urine, is critical for the diagnosis of early stage albuminuria, one of the major complications in nephropathy. The aim of this study was to investigate possible similarities or differences in albumin concentration and the presence of tyrosine in urine from a population of healthy and microalbuminuria dependent women.

Keywords: tyrosine, albumin, microalbuminuria, urine, UV-vis

---

[*] Corresponding Author Email: dbartusik-aebisher@ur.edu.pl.

INTRODUCTION TO MICROALBUMINURIA AND TYROSIN IN URINE

Normal human urine contains very small quantities of albumin. The appearance of larger amounts of albumin in urine is a sign of kidney disease and is also recognized as an independent risk factor for cardiovascular disease. Microalbuminuria is defined as a nephropathology linked with diabetic disease (Dabla 2010). The rate of albumin elimination in urine for healthy individuals is 30µg/day. Macroalbuminuria is diagnosed when the quantity of eliminated albumin is in the range of 100-200mg/L (De Cosmo et al. 2000). Microalbuminuria is diagnosed when albumin elimination is in the range of 30 and 100mg/L (De Cosmo et al. 2000). Numerous factors influence albumin elimination, e.g., excessive physical exercise, urinary infections, and an excess supply of protein in diet. The phenomenon of albumin excretion also has significant daily variability. Because of this variability, it is accepted that microalbuminuria is to be diagnosed when elimination levels indicating the disease state are found in at least two of three daily urine tests during 6 weeks.

The frequency of microalbuminuria occurrence in diabetes type I and type II patients is 20-40% (Roest et al. 2001; Monster et al. 2001). It appears that the measurement of albumin concentration is an indicator of increased risk, not only for peripheral nephropathy, but also for cardiovascular lesions, and development of arterial hypertension. In type I diabetes, microalbuminuria is treated as an indicator of development of chronic renal failure (Parving et al. 1982).

Moreover, the presence of tyrosine in urine of patients with microalbuminuria may be explained by endocrinal changes. A relationship between presence of 21-hydroxylase in blood and tyrosine in urine has been found (Parving et al. 1982; Viberti et al. 1982; Mogensen et al. 1987). This enzyme causes insulin hydrolysis resulting in release of large quantities of tyrosine. Elevated tyrosine levels in blood were found in menopausal women during measurements of hormone concentration (Parving et al. 1982; Viberti et al. 1982; Mogensen et al. 1987). Excess tyrosine is converted into glucose, and this explains the frequent occurrence of hyperglycaemia in cases of microalbuminuria (Parving et al. 1982; Viberti et al. 1982; Mogensen et al. 1987).

Interest in microalbuminuria increased when, at the end of the 1980s, it appeared that the condition is connected to vascular artheriosclerosis. Elevated albumin levels have since become one of the most important indicators of cardiovascular disorder in the diabetes-free population. Microalbuminuria, defined as a urinary albumin concentration of between 20-200mg/L, is an early predictor of diabetic nephropathy (Mogensen et al. 1983; Jarrett et al. 1984). In addition, microalbuminuria is known to be a marker for generalized vascular dysfunction e.g., a marker of cardiovascular morbidity and mortality, both in patients with diabetes mellitus and in the general population (Schmitz et al. 1988; Messent et al. 1992; Rossing et al. 1996; Allen et al. 2003; Damsgaard et al. 1992; Damsgaard et al. 1990, Borch-Johnsen et al. 1999; Hillege et al. 2001). The presence of both free and bound tyrosine in human urine has been known for some time, and has been verified in chromatographic analyses (Stein 1953).

Amino acid residues such as lysine, tyrosine, tryptophan and arginine are very important for functional properties of albumin (Block et al. 1953). Albumin contains 19 tyrosine residues which are considered to play an important role in its properties. Although albumin has been employed in the development of drug delivery systems through chemical modification, little attention has been paid to albumin functional groups for development of site specific protein delivery. Interest in the chemical composition and identification of urinary proteins has been stimulated by the work of Block et al. (Block et al. 1953).

Because many processes are regulated through albumin, it is important to gain insight into its functional mechanisms. Of the 360 grams of total albumin content in humans, 40% resides in the blood and 60% in extravascular fluids. About 40% of extravascular albumin is in muscle (Lichtman 1934). Albumin is responsible for the colloidal osmotic pressure of plasma, maintenance of blood volume, and supplies most of the acid/base buffering action of plasma proteins in extravascular fluids. Albumin binds and transports metabolites and drugs (Lichtman 1934 and Jain 1987). The transcapillary transport of albumin into extravascular spaces depends on the size and shape of the albumin molecule (Yuan et al. 1995).

Albumin plays several homeostatic roles, including maintenance of colloid osmotic pressure of blood and binding and transport of a wide range of compounds including drugs (Perl 1975; Nadal et al. 1996). Correlations between free amino acids in urine and biochemical changes have been determined.

An estimation of the L-tyrosine content of urine is of value in the assessment of hepatic disease and in the study of L-tyrosine metabolism (Jain 1987).

The albumin concentration was measured using an enzyme-linked immunosorbent assay and the urinary creatinine concentration was measured by standard laboratory methods. The albumin-creatinine ratio is at a level > 30mg/g for women with microalbuminuria (Mogensen et al. 1985; Mogensen 2000). There is a consensus that the diagnosis microalbuminuria requires two of three consecutive samples within this range (Mogensen et al. 1985).

ALBUMIN TO CREATININE RATIO MEASUREMENTS

The albumin-creatinine ratio is at a level > 30mg/g for women with microalbuminuria (Mogensen 2000). There is a consensus that the diagnosis microalbuminuria requires two of three consecutive samples within this range (Mogensen et al. 1985).

The albumin concentration was measured using an enzyme-linked immunosorbent assay and the urinary creatinine concentration was measured by standard laboratory methods.

The median value of the albumin-creatinine ratio in the 24 hour urine samples was calculated and patients were classified as having microabuminuria or nephropathy accordingly (Mogensen et al. 1985; Mogensen 2000).

A Review of Methodology: Sample Preparation for Determination of Tyrosine in Urine by UV-VIS

Mogensen et al. 1985; Mogensen 2000; Upstone 2000; Urine samples were prepared by adding basic lead acetate solution to urine and filtration through Whatman® filter paper. To the clear filtrate, potassium phosphate buffer was added in a volume equal to 1/6 of the sample volume and the clear supernatant was stored for further use. In the next step, arsenate borate buffer and phosphate buffer were mixed. Lyophylized *Crotalus adamanteus* venom (100mg) was stirred into 10mL of water and centrifuged. The clear supernatant (of Lyophylized *Crotalus adamanteus* venom) was added to a mixture of arsenate-borate buffer and phosphate buffer. Catalase (150,000U/mL) was added, the solution was diluted with phosphate buffer and mixed. The mixed reagent was allowed to stand followed by filtration through Whatman® 42 filter paper. The blank reagent was arsenate-borate buffer plus water diluted with phosphate buffer and filtrated. Basic lead acetate saturated solutions (containing $Pb(OH)_2$, and $Pb(CH_2CO_2)_2$ and potassium phosphate, the pH was adjusted to 6.5.

The concentration of phenylalanine in the plasma ultrafiolate (φ) is calculated as follows:

$$(\varphi) = (D_{308}T_{330} - D_{330}T_{308})/(P_{308}T_{330} - P_{330}T_{330}) \text{ (mg/100ml)}$$

where D_{308} is the absorbance of unknown at 330 nm with blank correction substracted; P_{308} is the calculated absorbance of phenylalanine solution 1.00mg at 308 nm with blank correction substracted; P_{308} is the calculated absorbance of phenylalanine solution, 100mg/100ml, at 308mµ; P_{330} is the calculated absorbance of phenyloalanine solution 1.00mg/100 ml, at 330mµ; T_{308} is the calculated absorbance of tyrosine solution 1.00mg /100ml at 308mµ; and T_{330} is the calculated absorbance of tyrosine solution 1.00mg/100ml at 330mµ. P_{308}, P_{330}, T_{308} are calculated using Beer's law from the absorbances of more concentrated solutions of phenylalanine and tyrosine measured at these two wavelengths. Since the ratio of T_{308} to T_{330} was practically constant at 1.75 the equation can be simplified to

$$(\varphi) = (D_{308} - 1.75 D_{330})/(P_{308} - 1.75 P_{330})$$

which can be used for calculating triphenyloalanine content of serum or plasma if slightly lower accuracy is acceptable. Tyrosine concentration in urine can be calculated from the results above.

Tyrosine concentration in urine = $(D_{330} - \varphi P_{330})/T_{330}$ mg/100 ml

Here (φ) is the concentration of phenyloalanine calculated as above. The concentration of phenylalanine in urine (Ψ U) is given by:

Phenylalanine concentration in urine = Ψ U = $(D_{308} - 1.9 D_{330} + 1.75 D_{350})/(P_{308} - 1.9 P_{330})$

where D_{350} is the absorbance of the unknown at 350 mµ, with blank correction subtracted.

Albumin concentration values, are measured on the basis of a change of maximum absorbance and acidic solution containing Comasie Brilliant Blue G-250 in the presence of protein from 465nm to 595nm.

The change in absorbance was proportional to protein concentration in solution.

Table 1. Absorbance found using pure solutions of phenylalanine and tyrosine

Wavelength (nm)	Phenylalanine	Tyrosine (nm)
308	0.610	0.942
330	0.053	0.531
350	0.02	0.040

THE PROSPECTIVE OF TYROSIN MEASUREMENTS DURING MICROALBUMINURA

Tyrosine is a main metabolite of the activity pathway of microalbuminura disease. Tyrosine is observed if higher albumin concentration detected in urine. The critical role of pH status in physiology and pathophysiology of living organisms is well recognized. Upon therapeutic intervention, the delivery, absorption and pharmacological effectiveness of drugs can be altered by changing the pH of their local environment. Therefore, spatially and temporarily addressed pH measurements in vivo are of considerable clinical relevance. The ionization of the phenolic group of L-tyrosine gave a pKa of 10.10. Its spectrophotometric behavior did not change on increasing the pH from 9.5 to 12.0. This demonstrates that most of the tyrosine residues in albumin are ionized at pH 5.5. The most reasonable explanation for this behavior is that the phenolic group is present in a hydrogen bonded structure in native albumin but upon making the solution strongly alkaline, this structure is abolished.

CONCLUSION

In conclusion, the determination of free amino acids in biological samples should be seen as a complementary tool in order to add further insight into the nature of the amino acids and protein metabolism disturbance occurring in uremia.

ACKNOWLEDGMENTS

Dorota Bartusik-Aebisher acknowledges support from the National Center of Science NCN (New drug delivery systems-MRI study, Grant OPUS-13 number 2017/25/B/ST4/02481).

REFERENCES

Allen, K. V., Walker, J. D. (2003) Microalbuminuria and mortality in long-duration type 1 diabetes. *Diabetes Care,* 26(8):2389 - 2391.

Borch-Johnsen, K., Feldt-Rasmussen, B., Strandgaard, S., Schroll, M., Jensen, J. S. (1999) Urinary albumin excretion. An independent predictor of ischemic heart disease. *Arteriosclerosis, Thrombosis and Vascular Biology,* 19(8):1992 - 1997.

Dabla, P. K. (2010) Renal function in diabetic nephropathy. *World Journal of Diabetes,* 1(2): 48 - 56.

Damsgaard, E. M., Froland, A., Jørgensen, O. D., Mogensen, C. E. (1992) Eight to nine year mortality in known non-insulin dependent diabetics and controls. *Kidney International,* 41(4):731 - 735.

Damsgaard, E. M., Froland, A., Jørgensen, O. D., Mogensen, C. E. (1990) Microalbuminuria as predictor of increased mortality in elderly people. *British Medical Journal,* 300(6720):297 - 300.

De Cosmo, A., Argilas, A., Misico, G., Thomas, S., Piras, G. P., Trevision, R., Perkin, P. C., Bacci, S., Zucaro, L., Morgaglione, M., Frittito, L., Pizutti, A., Tassi, V., Viberriti, V. G., Trischita, V. (2000) A PC-1 amino acid variant (K121Q) is associated with faster progression of renal disease in patients with type 1 diabetes and albuminuria. *Diabetes,* 49(3):521 - 524.

Hjime, A., Shuihi, N., Sauae, M. (1973) *Endocrinology Japan,* 20:97 - 101.

Hillege, H. L., Janssen, W. M., Bak, A. A., Diercks, G. F., Grobbee, D. E., Crijns, H. J., Van Gilst, W. H., De Zeeuw, D., De Jong, P. E., Prevend Study Group (2001) Microalbuminuria is common, also in a nondiabetic, nonhypertensive population, and an independent indicator of cardiovascular risk factors and cardiovascular morbidity. *Journal of Internal Medicine,* 249(6):519 - 526.

Jain, R. K. (1987) Transport of molecules in the tumour interstitium: a review. *Cancer Research,* 47(12):3039 - 3051.

Jarrett, R. J., Viberti, G. C., Argyropoulos, A., Hill, R. D., Mahmud, U., Murrells, T. J. (1984) Microalbuminuria predicts mortality in non-insulin-dependent diabetics. *Diabetic Medicine,* 1(1):17 - 19.

Lichtman, S. S. (1934) Origin and significance of tyrosinuria in disease of the liver. *Archives in Internal Medicine,* 53(5):680 - 688.

Messent, J. W., Elliott, T. G., Hill, R. D., Jarrett, R. J., Keen, H., Viberti, G. C. (1992) Prognostic significance of microalbuminuria in insulin-dependent diabetes mellitus: a twenty-three year follow-up study. *Kidney International,* 41(4):836 - 839.

Mogensen, C. E. (1987) Microalbuminuria as a predictor of clinical diabetic nephropathy. *Kidney International,* 31(2):673 - 689.

Mogensen, C. E., Christensen, C. K., Vittinghus, E. (1983) The stages in diabetic renal disease. With emphasis on the stage of incipient diabetic nephropathy. *Diabetes,* 32(Suppl. 2):64 - 78.

Mogensen, C. E., Chachati, A., Christensen, C. K., Close, C. F., Deckert, T., Hommekl, E., Kastrup, J., Lefebvre, P., Mathiesen, E. R., Feldt – Rasmussen, B., Schmitz, A., Vibereti, G. C. (1985) Microalbuminuria - An early marker of renal involvement in diabetes. *Uremia Investigation,* 9(2):85 - 95.

Monster, T. B. M., Janssen, W. M. T., Wilbert, M. T., de Jong, P. E., de Jong-van den Berg, L. T. W., Lolkje, T. W. (2001) Oral Contraceptive Use and Hormone Replacement Therapy Are Associated with Microalbuminuria. *Archives of Internal Medicine,* 161(16):2000 - 2005.

Nadal, A., Fuentes, E., Pastor, J., McNaughton, P. A. (1996) Albumin stimulates the uptake of calcium into subcellural stores in rat cortical astrocytes. *Journal of Physiology,* 492(Pt 3):737 - 750.

Parving, H. H., Oxenboll, B., Svendsen, P. A., Christiansen, J. S., Andersen, A. R. (1982) Early detection of patients at risk of developing diabetic nephropathy: a longitudinal study of urinary albumin excretion. *Acta Endocrinology,* 100(4):550 - 555.

Perl, W. (1975) Convection and permeation and albumin between plasma and interstitial. *Microvascular Research,* 10(1):83 - 94.

Roest, M., Banga, J. D., Janssen, W. M. T., Grobe, W. T., Sixma, J. P., Jong, E., DeZecuw, D., Van der Schaw, Y. T. (2001) Excessive urinary albumin levels are associated with future cardiovascular mortality in postmenopausal women. *Circulation,* 103(25):3057 - 3061.

Rossing, P., Hougaard, P., Borch-Johnsen, K., Parving, H. H. (1996) Predictors of mortality in insulin dependent diabetes: 10 year observational follow up study. *British Medical Journal,* 313(7060): 779 - 784.

Schmitz, A., Vaeth, M. (1988) Microalbuminuria: a major risk factor in noninsulindependent diabetes. A 10-year follow-up study of 503 patients. *Diabetic Medicine,* 5(2):126 - 134.

Stein, W. H. (1953) A chromatographic investigation of the amino acid constituents of normal urine. *Journal of Biological Chemistry,* 201:45.

Upstone, S. L. (2000) Ultraviolet/Visible light absorption spectrophotometry in clinical chemistry in *Encyclopedia of Analytical Chemistry* R. A. Meyers (Ed.) pp. 1699–1714 John Wiley & Sons Ltd, Chichester.

Viberti, G. C., Hill, R. D., Jarrett, R. J., Argyropoulos, A., Mahmud, U., Keen, H. (1982) Microalbuminuria as a predictor of clinical nephropathy in insulin-dependent diabetes mellitus. *Lancet,* 1(8287):1430 - 1432.

Yuan, F., Dellian, M., Fukumura, D., Leuning, M., Berk, D. A., Torchilin, V. P., Jain, R. K. (1995) Vascular permeability in a human tumour xenograft: molecular size dependence and cutoff size. *Cancer Research,* 55(17):3752 - 3756.

INDEX

A

absorption spectra, 23, 24, 43, 46, 54, 58
acid, 2, 3, 6, 7, 8, 9, 10, 16, 17, 23, 24, 31, 34, 38, 41, 52, 71, 75, 83, 93
acute respiratory distress syndrome, 2
aggregation, 5, 13, 82
aggregation process, 82
albumin, v, vi, vii, viii, ix, 1, 2, 3, 4, 5, 6, 7, 8, 9, 10, 11, 12, 13, 14, 15, 16, 17, 19, 20, 21, 22, 23, 27, 28, 29, 30, 31, 32, 33, 34, 35, 36, 37, 63, 69, 70, 71, 72, 74, 75, 76, 77, 79, 80, 81, 82, 83, 84, 85, 86, 87, 90, 91, 92, 93, 94, 96, 97, 98, 99
albuminuria, viii, x, 91, 98
amino, 2, 3, 5, 7, 15, 27, 34, 38, 47, 48, 49, 52, 59, 61, 71, 74, 81, 82, 94, 97, 98, 100
amino acid, 2, 3, 7, 15, 27, 34, 38, 47, 48, 49, 52, 59, 61, 71, 74, 81, 82, 94, 97, 98, 100
antimicrobial PDT, 40
antioxidant, v, vii, viii, 2, 6, 11, 12, 14, 16, 19, 21, 22, 23, 28, 29, 30, 31, 32, 33, 34, 35, 36, 80, 83, 86
antioxidant properties, v, vii, viii, 16, 19, 21, 22, 23, 29, 34, 35

antitumor, 63, 66, 68
antitumor agent, 68
aromatic hydrocarbons, 62
arterial hypertension, 92
ascorbic acid, 22, 25, 31, 33
association constant, ix, 38, 43, 58

B

bacteria, ix, 38, 61
bilirubin, vii, viii, 1, 10, 14, 16, 20, 30, 31
binding interaction, 38, 39, 43, 44, 46, 54, 55, 58, 74, 87
biochemistry, 74
biocompatibility, 84
biological activity, 82
biological consequences, 59
biological samples, 97
biological systems, 24
biomarkers, 21, 23, 30
biomolecules, 42, 63, 86
biotechnology, 74
bladder cancer, 70, 71, 75
blood, vii, viii, 1, 2, 4, 6, 12, 20, 23, 30, 34, 81, 82, 87, 88, 92, 93, 94
blood plasma, vii, viii, 1, 2, 12, 24, 30, 34

blood stream, 4
blood vessels, 4
bone marrow, 4
brain, 73, 75, 80

C

cancer, ix, 38, 39, 40, 42, 60, 70, 73, 74, 81, 84, 85, 89
cancer cells, ix, 38, 40, 42
cancer therapy, 39
cardiopulmonary bypass, 2
cardiovascular disease, 92
cardiovascular morbidity, 93, 98
catalytic properties, 23, 28
cell line, 60
cell surface, 70
chemical, 8, 11, 21, 22, 26, 80, 84, 90, 93
chemotherapeutic agent, 85
chemotherapy, 60, 88
chronic diseases, 15
chronic kidney disease, 36
chronic renal failure, 92
cortisol, 4, 20
cysteine, 4, 6, 7, 8, 9, 22, 25, 28, 29, 31, 41, 62, 87
cytotoxicity, 5, 63, 66

D

decomposition, 50, 59, 61, 65, 67, 68
derivatives, 36, 59, 60, 68, 82, 87, 88
detection, 15, 34, 59, 77, 81, 83, 99
detoxification, 2, 8, 31
diabetes, 5, 13, 23, 29, 34, 81, 92, 93, 98, 99, 100
diabetic nephropathy, 34, 35, 93, 98, 99
diabetic patients, 5, 22
dialysis, 23, 24, 33, 35, 80
diseases, 2, 21, 22, 28, 30
distribution, 21, 24, 73

DNA, 11, 60, 62, 63, 66, 67
DNA damage, 60, 67
drug activity, 79
drug delivery, viii, ix, 11, 27, 73, 79, 86, 93, 97
drug discovery, 60
drug release, 89
drugs, vii, viii, 5, 9, 10, 19, 20, 21, 80, 83, 84, 85, 93, 94, 97

E

electron, vii, viii, 12, 37, 38, 40, 42, 47, 48, 50, 52, 54, 60, 61, 63, 64, 66, 67, 68
electron microscopy, 12
electron transfer, 52
electron transfer, vii, viii, 37, 38, 40, 42, 47, 50, 52, 54, 60, 61, 63, 64, 66, 67, 68
electrostatic interaction, 46
endothelial cells, 3
end-stage renal disease, 25, 32
energy, 26, 40, 41, 42, 47, 64, 67
energy transfer, 26, 41, 47, 64, 67
environment, 59, 80, 97
enzymatic activity, 7, 11
enzyme, 64, 67, 92, 94
enzyme-linked immunosorbent assay, 94
excitation, 41, 55, 56, 57, 60, 67
excretion, 5, 15, 16, 92, 98, 99
exposure, 9, 10, 17
extraction, 48

F

fatty acids, vii, viii, 1, 7, 10, 12, 20, 23
fibromyalgia, 23, 25, 35
fluid, 20, 21, 22
fluorescence, vii, viii, 12, 26, 37, 47, 48, 49, 51, 53, 54, 55, 56, 59, 82
fluorescence quantum yield, 47, 56

fluorescence resonance energy transfer (FRET), 47, 48, 55. 56, 58
fluorinated drug, viii, ix, 79, 80, 83, 84
fluorine, 81, 82, 83, 84, 85, 89
folic acid, 11, 12, 65, 67, 68, 72

G

gadolinium, 71, 75
gastrointestinal tract, 5
glucose, 5, 22, 24, 30, 92
glutamic acid, 3
glutamine, 26
glutathione, 6, 8, 82, 85
glycosylation, 14, 16, 75

H

heart disease, 98
heart failure, 5, 22, 30
hemodialysis, 28, 33
hemoglobin, 5, 9
hemorrhage, 2
hepatitis, 35, 72, 77
hepatitis e, 77
hepatocellular carcinoma, 77
histidine, 9, 41, 62
homocysteine, 82, 84, 85, 87
human, vii, viii, ix, 1, 2, 3, 9, 11, 12, 13, 14, 15, 16, 17, 20, 22, 24, 27, 28, 29, 30, 31, 32, 33, 34, 35, 36, 59, 60, 63, 64, 65, 67, 68, 70, 71, 73, 74, 75, 76, 77, 80, 81, 82, 83, 84, 85, 86, 87, 88, 89, 90, 92, 93, 100
human body, 2, 20
hybridization, 36
hydrogen, 6, 8, 11, 40, 97
hydrogen peroxide, 6, 8, 11, 40
hydrolysis, 5, 7, 8, 15, 92
hydroperoxides, 59, 61
hydrophobic interaction, 46
hydrophobicity, 7

hydroxyl, 28, 40
hyperalbuminemia, 1, 2, 13
hyperglycaemia, 93
hypoalbuminemia, 13
hypoxia, 62, 71, 75

I

in vitro, 15, 22, 33, 34
in vivo, vii, viii, ix, 14, 19, 23, 29, 69, 73, 75, 80, 84, 97
infrared spectroscopy, 82
inhibition, 50, 52, 54
inhibitor, 8, 85, 90
insulin, 4, 92, 98, 99, 100
insulin dependent diabetes, 100
iron, 7, 9, 28, 70, 71, 77
irradiation, ix, 32, 38, 40, 56, 57, 73
ischemia, 2, 22, 23, 25, 30, 35

K

kidney, 5, 92

L

ligand, 7, 9, 13, 14, 15, 16, 24, 32, 62, 80, 85
light, vii, viii, 37, 40, 42, 43, 47, 48, 55, 57, 58, 60, 61, 63, 66, 100
liquid chromatography, 49
liver, 2, 4, 13, 30, 87, 99
liver disease, 2
liver failure, 2
low-density lipoprotein, 4, 26, 30
luminescence, 61
lysine, 3, 5, 7, 8, 10, 93

M

macromolecules, 80, 88
magnetic resonance imaging (MRI), 11, 27, 69, 70, 71, 72, 73, 74, 75, 76, 77, 84, 85, 86, 87, 97
magnetic resonance spectroscopy, 70, 87
mammalian brain, 77
mass spectrometry, 62, 75
matrix metalloproteinase, 74, 87
medical, 7, 16, 40
medicine, 11, 75, 77
mellitus, 5, 24, 93, 99, 100
metabolism, 94, 97
metabolites, 9, 10, 23, 29, 67, 93
metal complexes, 26
metal ion, 9
metal nanoparticles, 66
mice, 35, 72, 76, 77, 81, 89
molecular mass, 3, 5, 9
molecular orbital, 41, 42
molecular oxygen, 61, 64
molecular weight, 9
molecules, vii, viii, 2, 8, 9, 10, 12, 20, 22, 24, 37, 39, 40, 41, 47, 48, 52, 54, 98
mortality, 2, 11, 33, 93, 98, 99, 100
mortality risk, 11

N

nanoparticles, 25, 27, 70, 71, 72, 75, 76, 77
nephropathy, viii, x, 23, 24, 32, 91, 92, 94, 100
nephrotic syndrome, 2, 22, 32
nitric oxide, 6, 9, 20
nitrogen, 4, 6, 11
nitroxide, 27, 29
nuclear magnetic resonance (NMR), 16, 71, 72, 75, 77, 79, 80, 81, 82, 83, 84, 85, 86, 87, 88, 89, 90
nucleic acid, 11

nucleoside analogs, 87
nucleotide sequence, 15
nucleotides, 3

O

optical fiber, 40
optical properties, 73
organic matter, 62
osmotic pressure, 25, 93, 94
oxidation, vii, viii, 6, 7, 22, 23, 28, 31, 32, 34, 37, 38, 40, 42, 50, 52, 54, 57, 59, 60, 61, 62, 64, 66
oxidative damage, 21, 22, 39, 40, 49, 65, 68
oxidative stress, 5, 6, 15, 22, 23, 24, 28, 35, 36, 38
oxide nanoparticles, 70, 72
oxygen, 4, 6, 11, 32, 38, 40, 41, 42, 52, 54, 59, 61, 62, 63, 65, 67
oxyradical scavenger, vii, viii, 19

P

peptide, 4, 59, 68, 88, 90
pH, 3, 10, 16, 22, 33, 55, 56, 76, 89, 95, 97
phenothiazine dyes, vii, ix, 38, 40, 61, 67
phosphate, 8, 54, 56, 95
phosphorus, 63, 64, 67, 68
photodynamic therapy (PDT), ix, 38, 60, 61, 75
photoirradiation, 40, 49, 56
photooxidation, 39, 42, 62
photosensitizers, vii, ix, 38, 40, 41, 42, 43, 44, 45, 46, 47, 51, 52, 54, 55, 56, 58, 60, 61
polysaccharide, 81, 87, 89
population, viii, x, 91, 93, 98
porphyrin(s), vii, ix, 38, 40, 46, 51, 53, 54, 62, 63, 65, 66, 67, 68
prostate cancer, 25, 30

Index

protein damage, 21, 38, 50, 52, 57, 61, 63, 64, 67, 68
protein oxidation, 52, 59, 61, 64, 67
protein structure, 3
proteins, 9, 20, 24, 27, 39, 46, 59, 80, 82, 90, 93

Q

quantum yields, 50, 51
quercetin, 31

R

radical formation, 82
radicals, 9, 11, 24, 29, 36, 40, 59
reactive oxygen, vii, viii, 8, 9, 22, 25, 27, 36, 37, 38, 40, 58, 59, 60
reactivity, 23, 25, 29, 34
refractive index, 48

S

serum, vii, viii, ix, 1, 2, 3, 4, 5, 6, 7, 9, 11, 12, 13, 14, 15, 16, 17, 21, 22, 23, 27, 28, 29, 30, 31, 32, 33, 34, 35, 36, 37, 38, 59, 63, 64, 65, 67, 68, 70, 71, 73, 74, 75, 76, 77, 79, 80, 81, 82, 83, 84, 85, 86, 87, 88, 89, 90, 96
serum albumin, vii, viii, 1, 2, 3, 5, 6, 7, 9, 11, 12, 13, 14, 15, 16, 17, 21, 22, 23, 27, 28, 29, 30, 31, 32, 33, 34, 35, 36, 37, 38, 59, 63, 64, 65, 67, 68, 70, 71, 73, 74, 75, 76, 77, 79, 80, 82, 83, 84, 85, 86, 87, 88, 89, 90
singlet oxygen, 38, 40, 59, 61, 62, 63, 65, 67
small intestine, 4
sodium, 20, 54, 56, 60, 81, 82
solubility, 21, 80, 81, 88

solution, ix, 26, 38, 50, 51, 53, 54, 55, 56, 57, 81, 95, 96, 97
solvents, 81
species, vii, viii, 4, 8, 20, 22, 25, 29, 36, 37, 38, 40, 54, 58, 59, 60
spectrophotometry, 100
spectroscopy, v, vii, ix, 12, 26, 33, 61, 66, 69, 70, 73, 80, 81, 82, 83, 84, 86, 90
structural changes, 29, 35
structural modifications, 7
superparamagnetic, 71, 77
surface modification, 86

T

temperature, 3, 46, 55, 58
temperature dependence, 46, 55, 58
thermodynamic parameters, 46, 55, 58
thyroxin, vii, viii, 1, 72, 74
transactions, 74, 76
transition metal ions, 12
treatment, ix, 2, 9, 21, 38, 40, 61, 70
trifluoroacetic acid, 81
triiodothyronine, 20
tryptophan, vii, viii, 3, 6, 7, 8, 37, 38, 41, 42, 47, 48, 49, 53, 55, 56, 58, 59, 61, 83, 93
tryptophan oxidation, 38
tumor(s), 62, 71, 72, 81, 84, 87
tyrosine, vi, viii, x, 6, 7, 8, 31, 41, 49, 62, 85, 90, 91, 92, 93, 94, 95, 96, 97

U

ultraviolet irradiation, vii, viii, 37
urine, viii, ix, 2, 5, 91, 92, 93, 94, 95, 96, 97, 100
UV-vis, 23, 24, 91, 95

V

vaccine, 89
van der Waals force, 46
vitamins, 20

X

xenografts, 60
x-ray diffraction (XRD), 47

Antibiotic Resistance: Causes and Risk Factors, Mechanisms and Alternatives

Editors: Adriel R. Bonilla and Kaden P. Muniz

Series: Pharmacology - Research, Safety Testing and Regulation

Book Description: This book addresses the concern that over the past few years, there has been a major rise in resistance to antibiotics among gram-negative bacteria. New antibacterial drugs with novel modes of actions are urgently required in order to fight against infection.

Hardcover ISBN: 978-1-60741-623-4
Retail Price: $265

Medicinal Plants and Sustainable Development

Editor: Chandra Prakash Kala (Indian Institute of Forest Management, Madhya Pradesh, India)

Series: Pharmacology - Research, Safety Testing and Regulation

Book Description: This book deals with multidisciplinary approach and contains information on different aspects of medicinal plants, which can be used as guiding tools.

Hardcover ISBN: 978-1-61761-942-7
Retail Price: $139

Pharmaceutical Innovation: Challenges and Competitors

Editors: Tomas E. Sanchez and Charles L. Scott

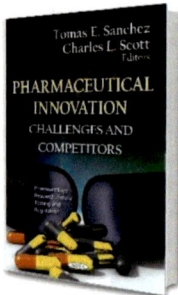

Series: Pharmacology - Research, Safety Testing and Regulation

Book Description: This book examines the challenges associated with striking the proper balance between lower cost drugs and maintaining an innovative domestic pharmaceutical sector.

Hardcover ISBN: 978-1-62257-068-3
Retail Price: $130

Human Serum Albumin (HSA): Functional Structure, Synthesis and Therapeutic Uses

Editor: Travis Stokes

Series: Protein Biochemistry, Synthesis, Structure and Cellular Functions

Book Description: This book provides an overview of the expanding field of preclinical and clinical applications and developments that use albumin as a carrier of drug delivery systems.

Hardcover ISBN: 978-1-63482-963-2
Retail Price: $230